Cedric Duvinage

Referees in Sports Contests

GABLER RESEARCH

Management, Organisation und ökonomische Analyse, Band 14

Herausgegeben von Professor Dr. Peter-J. Jost
WHU – Otto Beisheim School of Management, Vallendar

In den vergangenen zwei Jahrzehnten hat sich ein neuer mikroökonomischer Ansatz entwickelt, der nicht wie die traditionelle neoklassische Analyse auf den Marktbereich beschränkt ist, sondern der grundsätzlich für die Analyse sozialer Interaktionssituationen geeignet ist. Informationsökonomie, Spieltheorie, experimentelle Studien, Neue Institutionenökonomie und Ökonomische Psychologie sind wichtige Bausteine dieses ökonomischen Ansatzes.

Ziel der Schriftenreihe ist die Anwendung und Weiterentwicklung dieses Ansatzes auf betriebswirtschaftliche Fragestellungen. Gegenstand der Untersuchungen sind die unterschiedlichsten unternehmensinternen Probleme aus den Bereichen Finanzierung, Organisation und Strategisches Management. Die Reihe soll so zu einer mikroökonomischen Fundierung des Faches beitragen.

Cedric Duvinage

Referees in Sports Contests

Their Economic Role
and the Problem of Corruption
in Professional German Sports Leagues

With a Foreword by Prof. Dr. Peter-J. Jost

GABLER

RESEARCH

Bibliographic information published by the Deutsche Nationalbibliothek
The Deutsche Nationalbibliothek lists this publication in the Deutsche Nationalbibliografie;
detailed bibliographic data are available in the Internet at http://dnb.d-nb.de.

Dissertation WHU – Otto-Beisheim School of Management, 2011

1st Edition 2012

All rights reserved
© Gabler Verlag | Springer Fachmedien Wiesbaden GmbH 2012

Editorial Office: Marta Grabowski | Stefanie Loyal

Gabler is a brand of Springer Fachmedien.
Springer Fachmedien is part of Springer Science+Business Media.
www.gabler.de

Coverdesign: KünkelLopka Medienentwicklung, Heidelberg
Printed on acid-free paper

ISBN 978-3-8349-3526-7

To my brother and my parents

Foreword

Over the past few years, the role of the referee has been criticized considerably in professional German sports leagues: Mistaken calls that substantially influence the outcome of games are daily fare, and even deliberate game manipulations in favor of one of the competing teams or individual competitors have increasingly become the basis of public discussions. The scandal involving the former soccer referee Robert Hoyzer, who received considerable kickbacks for purposely manipulating the outcomes of professional soccer games in 2005, leading to a permanent ban from supervising games by the DFB, is only one of many examples that could be mentioned here.

If we consider the criticism of the role of the referee in a more differentiated way, two contextual interrogations arise: On the one side, mistaken calls in sports are based on an inherent misjudgment of game situations - either because of distortions of the human intake and processing of information or because of a deficient technical or personal support of the referee. The fundamental question here concerns the meaning of the referee with respect to the quality of the game: To what extent does the contribution of the referee foster a team's legitimate actions in a game on the one hand, and to what extent does it contain a team's irregular activities on the other hand? On the other side, corrupt delinquencies are based on a conscious error of judgment of game situations by the referee. In the course of this, the question of an efficient incentivization mechanism by sports associations is of prior interest: To what extent can changes in the rules of the game or monetary and immaterial incentives reduce or prevent a referee's propensity for corruption?

Cedric Duvinage's work addresses exactly these two questions: His objective is to provide answers to these questions in a game theoretical framework and to derive recommendations therefrom. From a scientific perspective, this intention is downright ambitious and takes up an academic void: The few existing studies that address the role of referees in sports are all of empirical or experimental nature; a theoretical elaboration of this topic is still missing.

Mr. Duvinage's work at hand excels to do exactly that in an outstanding way. However, not only theoretically interested readers but also those who are interested in sports and in the inferences for the improvement of the quality of referees as well as for a successful corruption prevention benefit from his work. I wish this work a respectively wide reception.

Peter-J. Jost

Acknowledgements

No matter whether at the professional or at the amateur level, all enthusiastic athletes and sports fans sooner or later make their own experiences with referees; certainly some good, but probably also some bad. Loving the game as a small forward on the basketball court over the course of many years, I have surely made mine.

However, while the high degree of sports commercialization has immensely increased the economic importance of sports referees, professional German sports leagues seem to have developed an evermore appearing corruption problem taking the occasional irritation about seemingly incompetent referees to a whole new level. For this reason, the study of sports referees deserves the attention of academic and economic research.

My particular personal relation to this topic as well as my interest in the well-being of the sports society surely boosted my motivation for intrinsic research in this area. But the completion of this project would not have been possible without the support of a special group of individuals.

Foremost, I would like to thank my academic advisor and dissertation supervisor, Prof. Dr. Peter-J. Jost, who showed tremendous interest in my research through countless discussions to a cup of tea, providing me with many helpful impulses and ideas. Furthermore, I would like to thank Prof.

Dr. Jürgen Weigand for his insightful and helpful comments as well as for taking on the task of the second supervisor.

I am also very thankful for the great support of all doctorate colleagues and friends without whom my time as a doctorate student by far would not have been as much fun and as enjoyable as it was. I especially thank all of those who took their time to share their own experiences with and as referees with me. The development of this dissertation irrefutably benefitted from these insightful chats and discussions.

Finally and most importantly, I would like to show my grand appreciation for the unconditional support of my family. I especially thank my greatest supporters and sponsors, my brother Christopher as well as my parents Peter and Angela, who I dedicate this thesis to.

Cedric Duvinage

Contents

List of Abbreviations

ad	to
Art.	Article
BC	Before Christ
BGB	Bürgerliches Gesetzbuch (German Civil Code)
BGH	Bundesgerichtshof (Supreme Court)
cf.	confer (compare)
DBB	Deutscher Basketball Bund (German Basketball Federation)
DBL	Deutsche Basketball Liga (German Basketball League)
DFB	Deutscher Fussball Bund (German Soccer Federation)
DFL	Deutsche Fussball Liga (German Soccer League)
DHB	Deutscher Handball Bund (German Handball Federation)
DHL	Deutsche Handball Liga (German Handball League)
DPA	Deutsche Presse Agentur (German Press Agency)
etc.	et cetera (and other things)
et al.	et alii/aliae (and others)
e.g.	exempli gratia (for example)
et seq.	et sequentes (and the following)

FA	Football Association
FIFA	Féderation Internationale de Football Association (International Football Federation)
FIVB	Féderation Internationale de Volleyball (International Volleyball Federation)
GG	Grundgesetz (Basic Constitutional Law)
i.e.	id est (that is)
IHF	International Handball Federation
LHS	Left Hand Side
Marg. No.	Marginal Note
NBA	National Basketball Association
p.	Page
para.	Paragraph
pp.	Pages
R&D	Research and Development
RHS	Right Hand Side
RuVO	Rechts- und Verfahrensordnung (Rules of Law and Procedure)
StGB	Strafgesetzbuch (Criminal Code)
StPO	Strafprozessordnung (Code of Criminal Procedure)
UEFA	Union of European Football Associations
WADA	World Anti-Doping Agency
ZPO	Zivilprozessordnung (Code of Civil Procedure)

Chapter 1

Introduction

Heaven and Hell decide to fight their differences by organizing a soccer match. Before the game, God proudly brags: "We've got all the star players!" Yet, the Devil counters: "So what? We've got the referees!"[1]

1.1 Motivation

With the increasing commercialization of professional sports and the rising public enthusiasm about various national and international sports events, professional sports referees are increasingly becoming the target of public scrutiny and criticism. Not long ago, during the round of sixteen match of the 2010 FIFA World Cup, England versus Germany, the Uruguayan referee Jorge Larrionda disgruntled many English soccer fans by disallowing an irrefutable goal for England. At the same time, many German soccer fans celebrated Larrionda's mistake as the revenge for the legendary *"Wembley*

---

[1] Freely translated from Steinke, Bodmer and Tophoven (2010, p. 64).

Goal" during the World Cup Final in 1966.[2]

The historic soccer rivalry between England and Germany vividly exemplifies that referees are undeniably a critical part of the game, and that it is not unusual for their judgments to ultimately determine its outcome. Given the immense commerce involved in some types of sports, incompetencies by referees, independent of whether they are deliberate or not, can cause a serious economic damage to competitors. This poses the question of why sports associations started availing themselves of the institution of referees in the design of professional sports contests in the first place.

At first sight, the introduction of referees might simply be explained by the contention that there has to be an objective supervisor responsible for the enforcement of the rules to ensure that competitors righteously attain the occasionally large benefits from winning a contest. However, the economic reasoning behind the role of referees goes far beyond this point.

As any competitive athlete can probably approve, a referee can significantly influence the nature of a game. The reason being that competitors typically adjust their playing strategies to the performance of the referee in the pursuit of maximizing their probabilities of success. From this follows that sports associations can use the referee in connection with the pre-specified rules of the game as an intrinsic instrument of design, so as to maximize the demand for its sports events.

Yet, blatantly wrong decisions such as that of Larrionda often tend to make enthusiastic spectators question the impartiality of a supervising referee. Indeed, recent corruption scandals in the professional German Soccer League (DFL, hereinafter also referred to as the "Fussball Bundesliga") and the professional German Handball League (DHL, hereinafter also referred to as the "Handball Bundesliga") provided ample reasons for developing

---

[2]   For more information, see Gödecke (2010), Reschke and Knaack (2010) and Volkery (2010).

such mistrust. The most recent ones in Germany involve the soccer referee
Robert Hoyzer and the handball referees Frank Lemme and Bernd Ullrich.[3]
The assertion of handball functionaries that even the president of the International Handball Federation (IHF), Hassan Moustafa, is up to corrupt
mischief also does not promise any comfort in this regard;[4] neither do the
recent suspicions about 200 possibly manipulated European soccer games,
where apparently more than 30 of them took place in Germany.[5]

Even more striking are the testimonies of retired referees, who experi-
enced gross bribery at first hand, warning sports fans that the corruption of
referees seems to be much more widespread than previously expected. Many
referees can tell stories about dubious bribery attempts. Sometimes, already
long before the game referees get phone calls and are asked to prepare wish
lists for their hosts. Alcohol, food, women and, of course, money make
the most popular means of bribery.[6] However, although the recent media
coverage focused its eyes on the revelation of sports corruption in the German Soccer Federation (DFB) and the German Handball Federation (DHB),
Willy Bestgen, a retired German basketball referee, asserts: "*Bribery occurs
in all sports.*"[7]

The term "*sports corruption*" is a colloquial expression and can be interpreted in many ways.[8] However, because the focus of this dissertation is the
study of the sports referee, the term "*sports corruption*" shall, in the context

---

3 For more details on the scandal involving Robert Hoyzer, see Ahrens (2005). For
   more information on the scandal involving Frank Lemme and Bernd Ullrich, see
   DPA (2009a).
4 For more information, see Hellmuth and Ewers (2009) and Weinreich (2009).
5 DPA (2009b); see also Peer (2009).
6 Stern (2009)
7 Tuneke (2009)
8 Maennig (2005, p. 189), for example, distinguishes between "*competition corruption*", relating to delinquent activities to influence competition results, and "*management corruption*", relating to illicit activities regarding non-competition decisions,
   such as those concerning host venues for large sports events, allocation of rights (e.g.
   television rights), or the constructions of sports arenas, for instance.

of this thesis, merely be defined as the bribery of referees by a competitor in order to gain an advantage within the contest. The economic analysis herein will therefore neither consider betting fraud, nor the bribery of players as a way of manipulating a game, nor any delinquent activities by functionaries in the sports industry.[9]

For a long time, the sports industry tried to keep the issue of sports corruption under wraps because it feared the reaction of spectators and sponsors.[10] After all, the obliquity inherent in sports corruption damages the integrity and sociability of sports. By all means, the sports ethos rests on fundamental values and principles that we consider desirable in today's society. It praises fairness, respect and loyalty, and it teaches the principle of equal opportunities and that hard work is rewarded.[11]

Until now, the German sports industry has always been able to attract the public by living and practicing these values. However, unless the sports industry now reacts quickly to solve its evermore apparent corruption problem, it will have difficulty to do so in the future - especially in light of the unsatisfactory legal clarity regarding sports corruption in Germany. The general application of the present regulations of the German Criminal Code (StGB) to delinquent activities of sports corruption, as it is defined in this dissertation, seems to lack judicial justification. It is therefore particularly important for the continuing demand for professional sports contests in the future that German sports federations rapidly establish efficient anti-corruption mechanisms.

The dissertation is structured as follows.

---

[9]   Although we might be inclined to categorize doping as a form of sports corruption, it should be viewed as a separate form of cheating. A doping competitor tries to elevate his winning chances by increasing his own performance relative to his rivals, while the bribery of a referee increases the briber's probability of success by effectively reducing the relative performance of his rivals. (See Preston and Szymanski (2003))

[10]   Ashelm (2009)

[11]   For a discussion on the role of sports in today's society, see McCormack (1988), Sage (1990) and Kleiber and Roberts (1981).

1.2 Structure

Following this chapter, Chapter 2 will provide a short overview of the relevant existing economic literature and embed this dissertation therein. This will comprise the reference to three branches of economic literature, that is the economic literature on referees, on contests and on corruption.

Chapter 3 will briefly consider the ancient history of sports referees. In particular, it intends to raise the awareness that the institution of referees is not an invention of modern times. It will demonstrate how already the ancient Olympic Games made use of referee-like functionaries to control ancient sports contests.

Chapter 4 will then develop a game theoretical model explaining the role of referees in sports contests, initially by defining the necessary parameters and by outlining the basic assumptions of the model. In the framework of a strategic conflict, the contestants' pure Nash equilibrium strategies in a symmetric contest without a referee will be derived before an honest referee is introduced to study how the referee affects the contestants' strategies. With the help of these results, it will then be possible to determine the value a referee creates for sports associations. Afterwards, the effect of asymmetries between players on the quality of a contest and also on the value of a referee is examined. Finally, Chapter 4 will demonstrate how corrupt referees become a real economic threat to the sports industry by considering their effect on the quality of sports contests.

Chapter 5 intends to shed light onto the problem of sports corruption in Germany from a legal perspective to explain why sports federations are concurrently struggling so much in preventing the bribery of referees. In particular, it will reveal the apparent legal loophole regarding the prosecution of sports corruption under present German criminal law. Moreover, a draft bill put forth by the judiciary of the Free State of Bavaria, intended to increase the efficiency of the deterrence of sports corruption in Germany, will be presented.

Chapter 6 will then develop an incentive model to study, from the perspective of a sports association, the optimal incentivization mechanism to prevent sports corruption. Aside from deriving the incentives of a contestant to bribe the referee, on the one hand, and the motivation of a referee to accept a bribe, on the other hand, this model intends to highlight the interdependence between the contestant's and the referee's incentives as well as the importance of this interdependence for the optimal design of anti-corruption policies.

Chapter 7 will then use the results established in Chapter 6 to develop and economically evaluate some possible anti-corruption policies. This will also include an assessment of the economic usefulness and efficiency of the Bavarian draft bill.

Finally, Chapter 8 will summarize the main findings of this dissertation and provide some proposals for future research on sports referees.

Chapter 2

Literature Review

2.1 Economic Literature on Referees

Until fairly recently, the study of referees has not received much attention in the economic literature. Merely with the increasing exploitation of the commercial potential in professional sports, scholars have started increasing their awareness of the economic importance of referees. The main topic among the few existing studies on referees has been the question of whether referees have the tendency of being one-sidedly influenced by external social pressures. Several researchers conducted experimental and empirical studies to examine the incidence of a natural referee home bias in professional sports leagues.

Nevill et al. (2002), for instance, conducted an experiment to test the impact of crowd noise on a referee's decisions. Two groups of referees were asked to watch the same game, where one group watched the game with and the other without volume. Interestingly, the group that watched the game with volume sanctioned fewer fouls of the home team. Sutter and

Kocher (2004) empirically investigated the referees' propensity of awarding extra time in the Fussball Bundesliga. They concluded that referees tend to add more extra time when the home team is behind. Dohmen (2005) also empirically confirmed the existence of a referee bias in the Fussball Bundesliga, where the composition and the size of the crowd seem to determine the direction and the extent of the bias.[1]

Undoubtfully, all of the existing experimental and empirical studies provide useful insights regarding the existence and the determinants of (unconscious) referee biases in professional sports contests. Yet, the previous research on referees has little to contribute to the understanding of the actual role of referees in sports contests. Especially, theoretical approaches explaining the role of referees are still absent in the economic literature.

2.2   Contest Literature

The contest literature, however, provides a large amount of at least related research for this dissertation. The surveys by Nitzan (1994) and more recently by Konrad (2009) provide informative overviews of the contemporary research while offering the basic theoretical aspects developed in the literature.

The origin of contest theory traces back to the pioneering contributions to the rent-seeking theory by Tullock (1967, 1980), Krüger (1974) and Posner (1975). In particular, Tullock's (1980) formulation of an imperfectly discriminating contest success function, which gained its eminent popularity among scholars due to its analytical tractability, laid down the groundwork for further insightful research on rent-seeking contests. Skaperdas (1996), Kooreman and Shoonbeek (1997) and Clark and Riis (1998) axiomatized Tullock's contest success function and suggested some constructive varia-

---

[1]   For more studies on referee biases, see also Garicano et al. (2005), Buraimo et al. (2010), Dawson et al. (2007) and Distaso et al. (2008).

tions.

Aside from the Tullock contest, researchers such as Hillman and Riley (1989) and Baye et al. (1996) modeled rent-seeking contests in the form of perfectly discriminating tournaments. Contrary to the Tullock contest, perfectly discriminating all-pay auctions guarantee the contestant exerting the highest effort level to win the tournament. However, as Lazear and Rosen's (1981) work shows, random factors can additively be accounted for in perfectly discriminating tournaments so that the contestant with the highest effort is no longer guaranteed to win. This way, an auction tournament with noise becomes, similar to a Tullock contest, imperfectly discriminating. Note that the assumption of an imperfectly discriminating contest success function, allowing for random external factors to affect the outcome of the contest, is in most sports contests the more realistic approach. Therefore, this will also be the approach taken in the model developed in Chapter 4 of this dissertation.

Scholars have studied perfectly and imperfectly discriminating contests along various dimensions in areas such as internal labor market tournaments (e.g. Lazear and Rosen (1981), Rosen (1986)), R&D races (e.g. Loury (1979), Dasgupta and Stiglitz (1980), Fudenberg, Gilbert and Tirole (1983), Harris and Vickers (1985), Leininger (1991)), political campaigning and lobbying (e.g. Skaperdas and Grofman (1995), Epstein and Hefeker (2003), Arbatskaya and Mialon (2007)), litigation (e.g. Katz (1988), Farmer and Pecorino (1999), Baye et al. (2005)), and sports contests (Ehrenberg and Bognanno (1990), Szymanski (2003a, 2003b), Frick (2003)).

Most of the present research focused on the incentive effects of the prize structure, of the number of contestants, of information asymmetries as well as of asymmetries in abilities and prize valuations between contestants.

The incentive effects of the prize structure of a contest was, for instance, studied by Lazear and Rosen (1981). They use an imperfectly discriminating

all-pay auction to compare the efficiency of incentives of risk-neutral workers under a *"piece-rate"* payment structure with a *"rank-order"* payment
structure. Nalebuff and Stiglitz (1983) generalize Lazear and Rosen's (1981)
work by studying relative performance compensation schemes in an imper-
fect information environment. They show that payment based on relative
performance (i.e. the rank-order payment structure) is superior to payment
based on absolute performance (i.e. the piece-rate payment structure) in a
high uncertainty environment.

Szymanski and Valletti (2005) analyze the optimal reward structure in
an asymmetric three-person Tullock contest. They argue that, in a Tullock
contest with one strong player and two weak players, awarding a prize only
to the winner of the contest is suboptimal. This is because the two weak
contestants would only have low incentives to exert effort in equilibrium,
knowing that the stronger contestant is most likely to win anyway. This in
turn also causes the strong contestant to exert a low effort in equilibrium
because he anticipates that the weaker contestants behave this way. It is
therefore argued that one has to give the weak contestants something to
fight for. This will then also make the strong contestant exert a higher
effort level because by exerting only a low effort the strong contestant would
run the risk of losing the contest against one of the weak contestants now
exerting high effort levels. Thus, the provision of a second prize is required
to extract optimal effort levels from all of the contestants.

A contest where contestants have asymmetric (i.e. incomplete) information about some characteristics of their opponents has been considered by
Amann and Leininger (1996). They assume in the form of a Bayesian game
that the players' unobservable types are drawn from a commonly known
distribution before effort levels are chosen to prove the existence and the
uniqueness of a Bayesian equilibrium in an asymmetric all-pay auction between two players.

The effect of asymmetries between players is the primary focus of Hillman
and Riley's (1989) study. They consider in particular asymmetries in the
players' prize valuations to examine their effect on the contestants' rent-
dissipation in perfectly and imperfectly discriminating contests. Baye et
al. (1996) elaborate on Hillman and Riley's (1989) analysis by highlighting
the possibility of a continuum of asymmetric equilibria. Baik (1994) uses
a logit formulation of the Tullock contest success function to study effort
levels between two asymmetric players, with specific regard to their relative
abilities and prize valuations. Stein (2002) extends Baik's (1994) study on
asymmetric abilities and prize valuations to contests with more than two
contestants.[2] Asymmetries in prize valuations and playing abilities will also
be considered in this dissertation to study the effect of such asymmetries on
the value of the referee.

Contests, specifically in the context of sports, have been studied by Szymanski (2003b). He provides an informative survey on the optimal design
of sports contests by considering various design elements such as the prize
structure, the optimal number of competitors, player asymmetries and the
concern for competitive balance between contest participants in a Tullock
contest framework.

Most studies (including all of the studies selected above), however, focus on contests where only positive rent-seeking activities (i.e. productive
efforts) are feasible. The possibility for contestants to also invest into negative rent-seeking effort (i.e. sabotage effort), which will be of essence in the
economic model developed in Chapter 4, has only relatively recently gained
more attention. Therefore, only a few scholars have addressed this problem

---

[2]   Asymmetric abilities and/or asymmetric prize valuations are also discussed by Rosen
     (1986), Dixit (1987), Harbring et al. (2004), Lazear and Rosen (1981), O'Keeffe et al.
     (1984), Glazer and Hassin (1988), Nti (1999), Clark and Riis (2000), Moldovanu and
     Sela (2001), Chen (2003), Kräkel and Sliwka (2004) and Cornes and Hartley (2005).

so far.[3]

The first attempt to model sabotage in a contest theory framework was made by Lazear (1989) who shows in an imperfectly discriminating all-pay auction model that pay equality (i.e. wage compression) reduces uncooperative behavior in a firm, where workers are paid based on relative performance. The reason is that wage compression will reduce the competition between workers within a firm to become the highest ranked worker, as they will all get a more equal share of the pie. As a result, workers have lower incentives to sabotage their colleagues in order to gain the extra wage.

Konrad (2000) studied the relationship between the contestants' sabotage effort and the number of contestants by modelling a variant of the Tullock contest. His underlying argument is that in tournaments with more than two players sabotage effort exerted by one competitor has a negative externality. This is because the sabotage effort exerted by one contestant harming another competitor not only increases the probability of winning of the sabotaging competitor but also that of all other non-sabotaged competitors. If this externality becomes too large (i.e. when the number of competitors increases), an equilibrium in which no sabotage occurs may emerge. Consequently, he argues that sabotage is more likely to occur in tournaments with few contestants. This phenomenon is, however, not confirmed by Harbring and Irlenbusch's (2008) experimental study, concluding that the number of contestants does neither affect the contestants' productive effort nor their sabotage effort.

Chen (2003) looks at promotion tournaments between asymmetrically able individuals in organizations. He concluded that more able individuals will in fact have lower chances of being promoted because they are subject to

---

[3]   Kräkel (2005) also considers the possibility that helping rather than sabotaging the opponent may oddly be an optimal equilibrium strategy for one of the players. He argues that helping allows players to self-commit before the game, leading to a less competitive tournament and thus to a lower expense of costly effort.

more sabotage than less able workers. Harbring et al. (2004) apply a logit formulation of the Tullock contest with asymmetric as well as symmetric players. They experimentally confirm their theoretical hypothesis that the composition of the competitors is crucial for the sabotage level in contests.

Yet again, none of the existing studies consider the impact of a referee on the strategic choices of contestants. Instead, in imperfectly discriminating contests the referee has implicitly been handled as an external random factor. However, because a referee can play a crucial role in determining the contestants' probabilities of success and thereby in influencing their strategic choices, it is worth modelling the role of a referee in sports contests explicitly.

2.3   Corruption Literature

To study the problem of sports corruption in Germany, this dissertation also intends to draw on the economic literature on corruption. The literature on corruption is indirectly linked to the literature on contests, as it points to the problem of corruption as a negative side effect of tournaments. It deals to a large extent with corrupt incentives of individuals in a principal agent framework and the optimal institutional design for corruption prevention. Jain (2001) and Andvig and Fjeldstad (2001) summarize the theoretical and empirical work on corruption and identify further areas for future research in this field.

Given the indirect connection to the contest literature, it is not surprising that the origin of the literature on corruption can also at least partially be traced back to the contribution on rent-seeking by Krüger (1974). Additionally, the corruption literature builds on the fundamental contributions to the economics of crime and punishment by Becker (1968) and Stigler (1970), as well as the basic contributions to the agency theory by Jensen and Meckling (1976), Fama (1980), Fama and Jensen (1983a, 1983b), Tirole (1986) and Eisenhardt (1985, 1989). The pioneering contribution to the economics of corruption, however, was made by Rose-Ackermann (1975) considering the link between changes in the market structure and the incidence of corrupt behavior by public officers.

Economists distinguish between two types of corruption: (1) Private-to-private corruption (i.e. corruption within and between firms) and (2) private-to-public corruption (i.e. bureaucratic corruption of a governmental officer by a private party).

Bliss and Di Tella (1997), for example, study the relationship between bureaucratic corruption and the degree of competition in the bribers' market. They argue that corruption itself affects the degree of competition in a free-entry equilibrium, as it affects the return on investments and thereby the number of firms in the market. Thus, the degree of competition cannot be considered as an exogenous determinant of corruption. Andvig and Moene (1990) examine the link between the profitability of economically motivated bureaucratic corruption and the frequency of corruption and explain why we observe varying degrees of corruption across societies by considering both the supply of and the demand for corruption.

Tirole (1986) raises the question of whether and, if so, how coalitions in the hierarchy of organizations might emerge. This idea of collusion in hierarchies will also be attended to in the model developed in Chapter 6 to discuss the contestant's and the referee's incentives for corruption. Based on Tirole's (1986) conclusion that collusion in hierarchies do indeed matter because they reduce the efficiency of the vertical structure, Bac (1996) examines the optimal monitoring structures and incentive schemes that minimize the cost of achieving a target level of corruption in hierarchies. He shows that the flatter the hierarchy, the more can monitoring costs be reduced due to monitoring economies of scale. Yet, a flatter hierarchy is also likely to increase the risk of collusion among agents.

Building on Becker's (1968) work on crime and punishment, Sosa (2004)

shows in a simple model that higher wages, typically viewed as an anti-
corruption policy, may in fact fail to deter corruption. Thereby, he essentially provides a theoretical explanation for the empirical findings of Van
Rijckeghem and Weder (2001). Conversely, Goel and Rich (1989) indeed
find evidence that high earnings in the pubic sector relative to the private
sector lower the level of corruption. Besley and McLaren (1993) develop a
model to assess the optimal conditions for three different pay schemes for
tax officers in order to maximize government tax revenue.

Some scholars have also already studied the problem of corruption with
a specific focus on the sports society. For example, with the help of a simple cost-benefit analysis, Maennig (2002, 2005) derives a range of general
anti-corruption policy suggestions for the sports society, some of which will
be assessed with the help of the economic model developed in Chapter 6.
Preston and Szymanski (2003), analytically discuss three types of cheating in sporting contests (match fixing, sabotage and doping) and provides
a short model illustrating the incentives for corrupt behavior provided by
betting markets. Mason et al. (2006) discuss the problem of corruption as
an agency problem in the International Olympic Committee and provides
some propositions for discouraging corruption in sports organizations.

Chapter 3

The Ancient History of Sports Referees

The idea of having an objective third party supervising a sports contest is not an invention of modern times. Already the ancient Greeks instituted the occupation of so called *"Hellanodicae"* (hereinafter referred to as *"Hellanodics"*) who were responsible for the supervision and administration of the Olympic Games.[1]

The historic year officially dating the beginning of the ancient Olympic Games is 776 BC. However, historians agree that this does not correspond to the factual commencement of the Olympic Games but rather to the first year in which the name of an Olympic victor - that is the name of the stadium runner Koroibos of Elis - was officially recorded.[2] However, archaeological evidence suggests that athletic activities, probably in connection to religious

---

[1]   Barney (2004, p. xxx)
[2]   Ebert (1980, p. 9)

events, were carried out at the site of Olympia well before 776 BC.[3] Whether or not Hellanodics were already introduced during that time is uncertain. Yet, the supervision of the Olympic Games by Hellanodics is confirmed until at least the beginning of the sixth century BC.[4]

The persons to become Hellanodics were randomly drawn from a pool of worthy residential applicants. Only members of the noble class were allowed to carry out the honorary tasks of Hellanodics.[5] Yet, someone who wanted to be given this honor was required to make contributions in various forms. On top of relinquishing all kinds of payment for their efforts, Hellanodics were also expected to provide considerable financial support for the Olympics.[6]

A Hellanodic's tenure was limited to one year, although multiple appointments might have been possible over a lifetime.[7] Hellanodics were always appointed already a year before the beginning of the Olympic Games, but they already took on their responsibilities and tasks immediately after the appointment.[8] Their wide ranging duties required an extensive preparation. Thus, already ten months before the opening ceremony, Hellanodics were meticulously informed about their tasks and the rules of the contests, where all of them had to swear the oath of office.[9]

Before an athlete could participate in the Olympic Games, Hellanodics were to check the integrity of a participant. Most importantly, they had to check whether an athlete was a free Hellenic citizen because non-Hellenes and slaves were excluded from participation. In addition, Hellanodics divided athletes into different age groups and discretionarily denied the participation of those that did not seem to fulfill the high standard of the Olympic tradition

---

[3]   Barney (2004, p. xxvii)
[4]   Ebert (1980, p. 69)
[5]   Finley and Pleket (1976, p. 111)
[6]   Finley and Pleket (1976, p. 112), Ebert (1980, p. 69)
[7]   Ebert (1980, p. 69)
[8]   Mezö (1930, p. 52)
[9]   Ebert (1980, pp. 69 et seq.), Mezö (1930, p. 52)

of athletic excellence.[10]

Hellanodics also had to supervise the preparatory practices that all contestants had to attend for a month. They laid down a pre-specified training schedule and, which seems odd for modern athletes, they even scheduled trial contests between the participating contestants before the actual games began. Those contestants who were late to such practice sessions or refused to take part in them were excluded from participating in the contests altogether.[11]

Their main and probably most difficult of all tasks was, however, the supervision of the contests. Three Hellanodics were assigned to each competition to make sure that the contestants fought according to the rules.[12] Any violation or unfair conduct was to be penalized via imposition of monetary fines, exclusion from the contest or corporal punishments. Finally, Hellanodics were in charge of identifying the victor and administrating the list of winners. In the pursuit of fulfilling their tasks they had a staff of trumpeters, stewards, heralds and punishers available to assist them.[13]

Thus, the main responsibility of Hellanodics in ancient contests was, similar to that of modern referees, the enforcement of a pre-specified set of rules. Hellanodics were also not able to change the rules of the contests or influence the design and structure of the contest in any way. This would have been unreasonable given their short tenure.[14]

Surprisingly, however, Hellanodics were also allowed to participate in the contests for quite some time. Yet, after Troilos, serving as a Hellanodic, also won the horse races in 372 BC, and complaints about favoritism arose among the competition, it was decided that they were no longer allowed to

---

[10]   Barney (2004, p. xxx), Ebert (1980, p. 70), Mezö (1930, pp. 52 et seq.); see also
       Finley and Pleket (1976, pp. 113 et seq.).
[11]   Finley and Pleket (1976, p. 116), Ebert (1980, p. 70)
[12]   Ebert (1980, p. 72)
[13]   Ebert (1980, pp. 70 et seq.)
[14]   Finley and Pleket (1976, p. 112)

do so.[15]

If athletes thought that they were put at a disadvantage by one of the Hellanodics, they could file a complaint against his judgment at the Olympic Committee. However, even if the complaint was justified, it could not lead to an overruling of the decision. Instead, the Olympic Committee would only impose a monetary fine on the accused Hellanodic.[16] This means that even the idea of a referee's irrevocable, factual and instant decision making, as it is inherent in modern sports contests, was already manifested in ancient sports contests as well.

To some extent, the problem of corruption also seems to have been an issue in ancient history already. The first officially reported corruption case occurred in 388 BC, where the fist fighter Eupolos of Thessalia bribed three of his opponents so that they would let him win the contest. However, the aim of such corrupt agreements only used to be the success in the contest. Corruption in relation to betting was unusual.[17]

Oddly, it was also typical that, if an athlete was discovered to have attended to bribery in order to win the contest, he got to keep the title. However, the authorities certainly availed of other ways of punishment in order to penalize cheating contestants.[18]

Even though historians believe that occasionally one also tried to suborn Hellanodics, the bribery of Hellanodics appeared to be rather rare. Hellanodics were said to be deputies of Zeus so that their honesty and incorruptibility was the highest of all prerequisites to uphold the reputation of the Olympic Games.[19] Besides, Hellnodics must have been hard to bribe considering that they were all members of the noble class. Furthermore, the types of contests

---

[15]   Ebert (1980, p. 70)
[16]   Mezö (1930, p. 54), Ebert (1980, p. 72)
[17]   Finley and Pleket (1976, p. 117 et seq.), Ebert (1980, p. 71)
[18]   cf. Finley and Pleket (1976, p. 118)
[19]   Mezö (1930, p. 56), Ebert (1980, p. 70)

that took place in the ancient Olympic Games did not include any team contests.[20] This means that bribing the opponent was at least as effective in increasing one's probability of success, and, due to the presumed honesty of Hellanodics, the attached risk of discovery was probably expected to be lower.

In any case, it is undeniable that a variant of the sports referee, as we know him today, already existed during ancient sports contests. The societal prestige of Hellanodics, their range of tasks and their responsibilities may have exceeded that of a modern sports referee. However, modern sports referees certainly still share their principal responsibility, that is the enforcement of the rules.

---

[20] For a timeline and a list of the contests introduced in the ancient Olympic Games, see Mezö (1930, pp. 60 et seq.).

Chapter 4

The Role of Referees in Sports Contests

Evidently, the role of referees must be of significant importance. Even after almost three thousand years, referees continue to be a vital part of sports contests. But what exactly is the role of referees? Does it really only entail the enforcement of the rules of the game? The following provides a new explanation relying on purely economic reasoning.

4.1 The Model

As Neale (1964) pointed out in his discussion on the peculiarity of the professional sports industry, the market structure of the sports industry is anything but usual. While the structural details of sports leagues certainly differ across regions, it is fair to generalize that a joint firm (i.e. the sports association) made up by several members (i.e. the sports clubs) sells a joint product (i.e. the game) produced by its competing members. The quality of the product is in essence determined by the quality of the interaction between the contestants, namely the displayed skills and performance.

Now, supposing that the demand (or the fan interest) increases with the game quality, the sports association is interested in optimally designing its sports contests in order to maximize the quality of the product.[1] Thus, via the contest design, including the prescription of the rules subject to which the game is to be played, the sports association tries to provide the contestants with incentives to expend their best playing efforts.

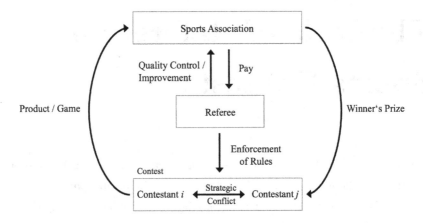

Figure 4.1: The underlying structure of the model

One way for the sports association to control and hopefully to improve the quality of the game is to hire a referee who is responsible for enforcing the rules of the game. This principal-supervisor-agent relationship, as illustrated in Figure 4.1, will be the basic structure underlying the model developed in this Chapter to analyze the effect of a referee on the strategic conflict between

---

[1] For more information on the optimal design of sports contests, see Ehrenberg and Bognanno (1990), Ross and Szymanski (2003), Szymanski (2003b) and Taylor and Trogdon (2002).

competitors.[2]

To keep the model simple, the game will be played between two risk-neutral contestants i and j, i.e. $i, j \in \{1, 2\}$ and $i \neq j$, where it is not important for this model whether these contestants are interpreted as teams or individual athletes. In any case, both competitors choose their effort levels simultaneously.

Each contestant can choose between two types of effort as instruments in order to increase his probability of success: (1) Productive playing effort (μ_i), where competitor i increases his probability of winning by increasing his own performance, and (2) sabotage effort (s_{ij}), where competitor i increases his winning probability by decreasing the opponent's (i.e. contestant j's) performance.

Thus, it is assumed that contestant i's performance/production function $q_i(\mu_i, s_{ji})$ is linearly increasing in his productive effort, μ_i, and decreasing in contestant j's sabotage effort s_{ji} on contestant i. Therefore, let contestant i's performance function in a game without a referee be given by

$$
\begin{aligned}
q_i^{hr}(\mu_i, s_{ji}) &= t_i^{\mu} \mu_i - t_j^{s} s_{ji} + +\varepsilon_i \\
&= y^{nr}(\mu_i, s_{ji}) + \varepsilon_i
\end{aligned}
\tag{4.1}
$$

where $t_i^{\mu} \geq 1$ is contestant i's talent for productive effort, μ_i contestant i's productive playing effort, $t_j^{s} \geq 1$ contestant j's talent to sabotage, and s_{ji} contestant j's sabotage effort on contestant i.[3] The error term ε_i, that could be interpreted as luck here, is an independently normally distributed random variable representing all other factors (except the referee) that are not in the control of the contestants but nevertheless might influence the outcome of

---

[2] For more information on the game theoretical analysis of strategic conflicts, see Jost (1999, pp. 54 et seq.).

[3] Harbring and Irlenbusch (2008) use a similar production function, in which productive effort and sabotage activity affect the overall performance linearly.

the game. The ability parameters t_i^μ and t_j^s could also be viewed as efficiency parameters because they measure the effectiveness with which each type of effort affects the overall performance of each contestant. The superscript nr in equation (4.1) indicates contestant i's performance function in a game with *no referee*.

With regard to the game quality, it will be assumed that the average quality of a game is increasing in the contestants' productive efforts μ_i and μ_j, and decreasing in their sabotage efforts s_{ij} and s_{ji}. Thus, let the average quality of a game between two contestants be determined by:

$$\phi = \frac{\frac{\mu_i}{s_{ji}} + \frac{\mu_j}{s_{ij}}}{2} \tag{4.2}$$

The idea behind the average quality function (4.2) is that when the contestants already extensively choose to sabotage the opponent, while exerting low productive effort, an additional unit of sabotage lowers the already poor quality of the game by less than an additional unit of productive effort could raise the quality. Similarly, when contestants exert a high level of productive effort and a low level of sabotage effort, an additional unit of sabotage lowers the game quality by more than an extra unit of productive effort would raise the quality of the already high quality game.[4]

As mentioned earlier, the sports association can hire a referee, ideally, in order to improve the game quality. The referee's task will then be to correctly identify and penalize sabotage effort according to the rules of the game. However, a referee occasionally makes mistakes. This includes two types of mistakes:

1. Type 1 error: The referee mistakenly calls a legitimate productive

---

[4]   Note that $\frac{\partial \phi}{\partial \mu_i} = \frac{1}{2s_{ji}}$ and $\frac{\partial \phi}{\partial s_{ji}} = -\frac{\mu_i}{2s_{ji}^2}$ so that $\left|\frac{\partial \phi}{\partial \mu_i}\right| = \left|\frac{\partial \phi}{\partial s_{ji}}\right|$ when $\mu_i = s_{ji}$, $\left|\frac{\partial \phi}{\partial \mu_i}\right| < \left|\frac{\partial \phi}{\partial s_{ji}}\right|$ when $\mu_i > s_{ji}$ and $\left|\frac{\partial \phi}{\partial \mu_i}\right| > \left|\frac{\partial \phi}{\partial s_{ji}}\right|$ when $\mu_i < s_{ji}$.

effort of contestant i as illegitimate with probability $m_i \in [0,1]$.

2. Type 2 error: The referee fails to convict a sabotage activity of contestant j on contestant i with probability $\overline{m}_i \in [0,1]$.

It will be assumed that $m_i < 1 - \overline{m}_i$, that is, it is more likely that the referee correctly penalizes sabotage effort than that he falsely sanctions productive effort.

Thus, supposing that the referee is honest (i.e. not corruptible) in the sense that m_i and $\overline{m}_i$ merely represent a referee's exogenously given unconscious errors to the detriment of competitor i, the performance function of contestant i transforms to

$$
\begin{aligned}
q_i^{hr}(\mu_i, s_{ji}) &= t_i^\mu \mu_i - m_i t_i^\mu \mu_i (1+F) - t_j^s s_{ji} + (1-\overline{m}_i) t_j^s s_{ji} F + \varepsilon_i \\
&= t_i^\mu \mu_i [1 - m_i(1+F)] - t_j^s s_{ji} [1 - (1-\overline{m}_i)F] + \varepsilon_i \\
&= y^{hr}(\mu_i, s_{ji}) + \varepsilon_i
\end{aligned}
\tag{4.3}
$$

where F is the marginal penalty awardable by the referee subject to the rules of the game. Assume, for simplicity, that the referee can impose only one type of penalty in a game so that the marginal penalty F is constant. The superscript hr indicates the case of a game with an *honest referee*.

Expression (4.3) exhibits that contestant i's performance increases with his own productive effort as long as the referee does not mistakenly penalize it. If, however, the referee convicts contestant i's productive effort as sabotage effort, contestant i not only loses his productive effort but is in addition also penalized for it with the imposed penalty F. Conversely, contestant i's performance decreases with every sabotage activity exerted by contestant j, no matter whether the referee calls it or not. If he does sanction it, however, contestant i at least benefits from the penalty imposed by the referee on contestant j. In other words, all sabotage activities by contestant j will reduce contestant i's performance. But if the referee correctly penalizes contestant j's sabotage effort, contestant i is at least compensated for it with F.

In order to ensure that the contestants expend positive productive effort in this model, the condition

$$\frac{\partial q_i^{hr}}{\partial \mu_i} = t_i^{\mu} \left[1 - m_i(1 + F) \right] > 0,$$

that is

$$\frac{1 - m_i}{m_i} > F, \tag{4.4}$$

must be assumed to be satisfied throughout the whole discussion.[5] Thus, given a marginal penalty F, condition (4.4) requires a referee's type 1 error to be low enough for the contestants to be willing to exert any productive effort at all.

As pointed out earlier, both contestants expend effort in order to improve their winning chances in such contest. Accordingly, the probability that contestant i wins the game will be given by

$$
\begin{aligned}
P(\mu_i, s_{ij}, \mu_j, s_{ji}) &= prob(q_i > q_j) \\
&= prob(y(\mu_i, s_{ji}) + \varepsilon_i > y(\mu_j, s_{ij}) + \varepsilon_j) \\
&= prob(y(\mu_i, s_{ji}) - y(\mu_j, s_{ij}) > \varepsilon_j - \varepsilon_i) \\
&= G\left[y(\mu_i, s_{ji}) - y(\mu_j, s_{ij}) \right]
\end{aligned}
\tag{4.5}
$$

where $G[\cdot]$ is the cumulative distribution function of the random composed variable $x := \varepsilon_j - \varepsilon_i$. The variables ε_j and ε_i are distributed according

---

[5]   Note that an equivalent condition for sabotage effort, i.e. $\frac{\partial q_i^{hr}}{\partial s_{ji}} = -t_j^s \left[1 - (1 - \overline{m}_i)F \right] < 0$, that is $\frac{1}{1 - \overline{m}_i} > F$, would be counter-productive because sabotage efforts between contestants would optimally be fully deterred to attain a high game quality. Hence, this is not a necessary condition for a contest to take place.

to a standard normal distribution, i.e. $\varepsilon_j \sim N(0, \sigma^2)$ and $\varepsilon_i \sim N(0, \sigma^2)$.[6]
Thus, the composed variable $x := \varepsilon_j - \varepsilon_i$ is also normally distributed, i.e.
$x \sim N(0, \sigma_x^2)$ where $\sigma_x = 2\sigma^2$.

It will further be assumed that investing into effort is increasingly costly.
Thus, let contestant i's cost function for legitimate effort be $c(\mu_i) = \frac{c_i}{2}\mu_i^2$ and
for sabotage $k(s_{ij}) = \frac{k_i}{2}s_{ij}^2$, where $c_i > 0$ and $k_i > 0$ to guarantee convexity
of the cost functions. The use of convex cost functions is explained by the fact
that each additional unit of effort in sports contests tends to be physically
more straining than the previous unit of effort.

Aside from incurring the costs of effort, the contestants also benefit from
participating in a contest. Each contestant attaches a certain value to the
winning prize, where for now let us assume that both contestants value the
prize equally as V. Thus, contestant i maximizes his expected payoff given
by:[7]

$$
\begin{aligned}
E\left[\pi_i\right] &= P(\mu_i, s_{ij}, \mu_j, s_{ji})V - c(\mu_i) - k(s_{ij}) \\
&= G\left[y(\mu_i, s_{ji}) - y(\mu_j, s_{ij})\right]V - \frac{c_i}{2}\mu_i^2 - \frac{k_i}{2}s_{ij}^2 \quad (4.6)
\end{aligned}
$$

Finally, assume for simplicity that throughout the whole analysis the
participation constraint, namely that the expected benefit from participating
in the contest is at least as great as the expected cost, is always satisfied for
both contestants.

---

[6]   The inclusion of the error terms ε_i and ε_j makes this auction model imperfectly
    discriminating. Lazear and Rosen (1981), Chen (2003), Kräkel and Sliwka (2004)
    and Kräkel (2006), for example, apply this type of success function, while Dixit
    (1987, p. 893) and Konrad (2009, pp. 37 et seq.) briefly discuss it as well.

[7]   For more information on the theory of utility maximization and rational decision
    making involving uncertain outcomes, see Jost (2000b, pp. 331 et seq.).

4.2 Symmetric Games

4.2.1 A Symmetric Game without a Referee

Today, a professional sports game without a referee, no matter whether in soccer, handball, basketball or in any other established type of sport, is hard to imagine. Yet, most sports were originally invented without incorporating a referee as a vital part of the game.[8] Hence, before introducing a referee into the game, consider first the players' equilibrium incentives in a game without a referee. The results established in this section will then serve as a benchmark for studying the effect of a referee on the players' incentives and on the quality of the game in the following sections.

In the case of a game without a referee, each contestant will choose his effort levels so as to maximize his expected payoff, i.e.

$$\max_{\mu_i, s_{ij}} E\left[\pi_i\right] = G\left[y^{nr}(\mu_i, s_{ji}) - y^{nr}(\mu_j, s_{ij})\right] V - \frac{c_i}{2}\mu_i^2 - \frac{k_i}{2}s_{ij}^2.$$

Recall from equation (4.1) that $y^{nr}(\mu_i, s_{ji}) = t_i^\mu \mu_i - t_j^s s_{ji}$. Now, taking the first order conditions of this optimization problem with respect to productive playing effort μ_i and sabotage effort s_{ij} yields the following optimal effort levels for contestant i:

$$\mu_{i_{nr}}^* = \frac{t_i^\mu g^{nr}\left[\cdot\right] V}{c_i}$$

$$s_{ij_{nr}}^* = \frac{t_i^s g^{nr}\left[\cdot\right] V}{k_i}$$

---

[8]   The soccer referee, for example, with his own discretionary decision-making power, as we know it today, was not introduced until 1889, which is twenty-six years after the foundation of the Football Association (FA) in London in 1863 (Huba (2007, pp. 21 et seq.)).

To simplify the notation here, $g^{nr}\left[\cdot\right] := g\left[y^{nr}(\mu_i, s_{ji}) - y^{nr}(\mu_j, s_{ij})\right]$.

If an ex-ante symmetric contest is assumed, the contestants are homogeneous in all characteristics including their playing talents, i.e. $t_i^\mu = t_j^\mu = t^\mu$ and $t_i^s = t_j^s = t^s$, and their marginal costs of effort, i.e. $c_i = c_j = c$ and $k_i = k_j = k$. From this ex-ante symmetry, here also leading to an ex-post symmetry in pure strategies, meaning that $\mu_i^* = \mu_j^* = \mu^*$ and $s_{ij}^* = s_{ji}^* = s^*$, we know that $g^{nr}\left[\cdot\right] = g\left[0\right]$. It follows that in a symmetric equilibrium the contestants' optimal strategies become:

$$\mu_{nr}^* = \frac{t^\mu g\left[0\right] V}{c} \tag{4.7}$$

$$s_{nr}^* = \frac{t^s g\left[0\right] V}{k} \tag{4.8}$$

As we would intuitively approve, equations (4.7) and (4.8) show that the optimal effort choices increase with the valuation of the winner's prize, the playing talent and the marginal winning probability. Conversely, the equilibrium effort levels decrease with the marginal cost of effort.

Referring to equation (4.2), the average game quality can now directly be derived as follows.

$$\phi_{nr} = \frac{\frac{\mu_{nr}^*}{s_{nr}^*} + \frac{\mu_{nr}^*}{s_{nr}^*}}{2} = \frac{\mu_{nr}^*}{s_{nr}^*} = \frac{t^\mu k}{t^s c} \tag{4.9}$$

Thus, the average quality of a game with no referee between two symmetric contestants only depends on the playing talents as well as the marginal costs for productive and sabotage effort. The average quality increases with the amount of productive playing talent in the game as well as the contestants' marginal cost of sabotage effort, while it decreases with the amount of sabotage talent as well as the contestants' marginal cost of productive effort. Note, in particular, that the quality of a game here is independent of the contestants' valuations of the winner's prize and the marginal winning probability.

4.2.2   A Symmetric Game with an Honest Referee

Having examined the contestants' incentives in a game without a referee, assume now that an honest referee is introduced into the game. If the referee is honest, the probability that the referee unconsciously makes a mistake to the detriment of contestant i must be equal to the probability that he unconsciously makes a mistake to the detriment of contestant j, so that $m_i = m_j = m$ and $\overline{m}_i = \overline{m}_j = \overline{m}$. The unconscious errors are exogenously determined and can not be instantly manipulated by the referee. Therefore, contestant i now optimizes his effort choices so as to

$$\max_{\mu_i, s_{ij}} E\left[\pi_i\right] = G\left[y^{hr}(\mu_i, s_{ji}) - y^{hr}(\mu_j, s_{ij})\right] V - \frac{c_i}{2}\mu_i^2 - \frac{k_i}{2}s_{ij}^2.$$

Recall from expression (4.3) now that in a game with an honest referee $y^{hr}(\mu_i, s_{ji}) = t_i^\mu \mu_i \left[1 - m_i(1 + F)\right] - t_j^s s_{ji} \left[1 - (1 - \overline{m}_i)F\right]$. If we assume ex-ante symmetry between the competitors once more, their optimal strategies for productive playing effort and sabotage effort in equilibrium are:[9]

$$\mu_{hr}^* \;=\; \frac{t^\mu \left[1 - m(1 + F)\right] g\left[0\right] V}{c} \tag{4.10}$$

$$s_{hr}^* \;=\; \frac{t^s \left[1 - (1 - \overline{m})F\right] g\left[0\right] V}{k} \tag{4.11}$$

Thus, as in the game without a referee, the equilibrium playing efforts increase in the valuation of the winner's prize, the playing talent, and the marginal probability of winning, while they decrease with the marginal cost of effort.

---

[9]   Remember, this implies that $g^{hr}\left[\cdot\right] := g\left[y^{hr}(\mu_i, s_{ji}) - y^{hr}(\mu_j, s_{ij})\right] = g\left[0\right]$.

However, in the game with a referee the equilibrium strategies for productive playing effort and sabotage effort additionally depend on the marginal awardable penalty as well as the referee's type 1 and type 2 errors respectively. We observe, on the one hand, that the higher the type 1 error m, the lower the productive effort levels exerted by the contestants. On the other hand, the higher the type 2 error $\overline{m}$, the higher the contestants' sabotage levels. This is because the referee's errors in essence influence the effectiveness of the contestants' strategies. The higher a referee's type 1 and type 2 error, the less effective a contestant's productive playing effort and the more effective a contestant's sabotage effort respectively in increasing the winning chances. This is summarized in Proposition 1.

Proposition 1 *A referee's errors manipulate the effectiveness of each type of effort exerted by the contestants. As a result, his type 1 error lowers the contestants' productive effort levels and his type 2 error increases their sabotage effort levels. At the same time, both productive and sabotage effort decrease with the marginal penalty awardable by the referee.*

Making use of equation (4.2) again, the average quality of a game with a referee is then given by

$$\phi_{hr} = \frac{\frac{\mu_{hr}^*}{s_{hr}^*} + \frac{\mu_{hr}^*}{s_{hr}^*}}{2} = \frac{\mu_{hr}^*}{s_{hr}^*} = \frac{t^\mu k \left[1 - m(1 + F)\right]}{t^s c \left[1 - (1 - \overline{m})F\right]}. \tag{4.12}$$

4.2.3   The Value of an Honest Referee

Now, a sports association might be interested in a referee's value, as it may be useful in determining a performance-based pay for referees. Notice from expression (4.12) that the average quality of a game with a referee increases with the performance of the referee, i.e.

$$\frac{\partial \phi_{hr}}{\partial m} < 0 \text{ and } \frac{\partial \phi_{hr}}{\partial \overline{m}} < 0. \tag{4.13}$$

This explains why sports federations typically only allow a selected group of the best referees available to supervise professional sports contests, and why such referees are generally required to attend a certain amount of training sessions per year.[10] Comparing equation (4.9) with (4.12), it therefore becomes apparent that the referee can actually increase the average quality of a game compared to the game quality without a referee if

$$\frac{[1 - m(1 + F)]}{[1 - (1 - \overline{m})F]} > 1, \tag{4.14}$$

or equivalently, if

$$F > \frac{m}{1 - m - \overline{m}}. \tag{4.15}$$

Proposition 2 *The referee increases the quality of a contest if $F > \frac{m}{1 - m - \overline{m}}$.*

In this way, a referee is an important instrument for sports federations in the optimal design of sports competitions, where the value of a referee can be readily depicted from expression (4.14). Only if $\frac{[1-m(1+F)]}{[1-(1-\overline{m})F]} - 1 > 0$, the referee improves the quality of the game. The value of the referee is therefore simply determined by the percentage growth in the game quality, as the referee is introduced into the game:

$$\frac{\phi_{hr}}{\phi_{nr}} - 1 = \frac{\frac{t^\mu k}{t^s c} \frac{[1-m(1+F)]}{[1-(1-\overline{m})F]}}{\frac{t^\mu k}{t^s c}} - 1 = \frac{[1 - m(1 + F)]}{[1 - (1 - \overline{m})F]} - 1 \tag{4.16}$$

Note that the value of the referee only depends on his type 1 and type 2 errors and the marginal penalty he is allowed to impose on wrongdoing

---

[10]   cf. DFB (2010a, Section 4), DBB (2008, Section 2), or DHB (2007, Section 4); DFB (2010a, Section 7(2)), DBB (2008, Section 9), or DHB (2007, Section 3)

contestants according to the rules of the game. It is independent of the marginal cost parameters c and k, and independent of the contestants' talents for productive effort and sabotage effort t^μ and t^s.[11]

Proposition 3 *The value of a referee is the percentage growth in the game quality as a result of his employment. It depends only on his type 1 and type 2 errors and the marginal penalty awardable by the referee subject to the rules of the game.*

4.3 Asymmetric Games

Having examined contests between symmetric contestants, we can now go on to study how asymmetries in prize valuations and playing talents/marginal costs of effort between players affect the results established above. Again, a game without a referee will be studied first in order to examine whether and, if so, how asymmetries between contestants affect the quality of the game in general. Afterwards, an honest referee will be introduced into the game in order to compute the value of a referee in an asymmetric game. The effect of asymmetries between players on the value of the referee is then determined via comparison of the value of the referee in an asymmetric game with that in a symmetric game.

4.3.1 An Asymmetric Game without a Referee

To assess the effect of asymmetries on the quality of a game, the various specific asymmetry parameters will be considered individually, using the case

---

[11] Here, the referee's type 1 and type 2 errors are assumed to be exogenously given. However, if one would endogenize these error terms, it would be inappropriate to argue that the value of a referee is independent of the sabotage talent t^s, although t^s does not directly appear in expression (4.16). The reason is that one must assume that the referee's performance is worsened as the contestants' talents for sabotage increase. For example, as competitors get better in hiding their sabotage activities or in misleading the referee, the referee's errors tend to increase, meaning that $\frac{\partial m}{\partial t^s} > 0$ and $\frac{\partial \overline{m}}{\partial t^s} > 0$. Such an anticipated effect of sabotage abilities on the value of a referee will be studied in Section 4.3.2.

of the symmetric contest, i.e. equations (4.7) and (4.8), as the benchmark.

Asymmetric Prize Valuations Consider first the possibility that the two competitors value the winner's prize differently. Let the variable $0 \leq \alpha < 1$ represent the asymmetry in prize valuations for the competing players, where $V_i = (1-\alpha)V$ and $V_j = (1+\alpha)V$.[12] So far, it was assumed that $\alpha = 0$. Now, we study the case where ceteris paribus $0 < \alpha < 1$. In doing so, players are still assumed to be symmetric in all other parameters, i.e. $t_i^\mu = t_j^\mu = t^\mu$, $t_i^s = t_j^s = t^s$, $c_i = c_j = c$ and $k_i = k_j = k$. This converts equations (4.7) and (4.8) to

$$\mu_{i_{nr}}^* = \frac{t^\mu g^{nr}[\cdot](1-\alpha)V}{c} \tag{4.17}$$

$$s_{ij_{nr}}^* = \frac{t^s g^{nr}[\cdot](1-\alpha)V}{k} \tag{4.18}$$

for contestant i's effort choices and to

$$\mu_{j_{nr}}^* = \frac{t^\mu g^{nr}[\cdot](1+\alpha)V}{c} \tag{4.19}$$

$$s_{ji_{nr}}^* = \frac{t^s g^{nr}[\cdot](1+\alpha)V}{k} \tag{4.20}$$

for contestant j's effort choices. Note that, contrary to the contests studied so far, $g^{nr}[\cdot] \neq g[0]$ in equilibrium. Due to the normal distribution of the composed variable $x := \varepsilon_j - \varepsilon_i$ around zero, it is in fact unambiguous that $g^{nr}[\cdot] < g[0]$. Expressions (4.17) to (4.20) imply that an asymmetry has two effects, a *direct* effect and an *indirect* effect.

The *direct* effect is that competitor j values the prize more than contestant i. This means that ex-ante contestant j has a higher expected benefit

---

[12] There are several different ways of modelling asymmetries in a contest model. The one used here was inferred from Szymanski (2003b, p. 1143).

from winning the contest than contestant i. Therefore, contestant j will be willing to exert more of both types of effort in equilibrium than contestant i in order to win the contest. This results in an increase in contestant j's equilibrium probability of success and an equivalent decrease in contestant i's probability of success.[13]

In turn, the *indirect* effect is initiated by the reduction in the marginal probability of winning for both contestants, as a result of the shift in the contestants' equilibrium probabilities of success. Being aware of this asymmetry and because effort is costly, contestant i maximizes his utility by indirectly lowering his effort choices. However, anticipating contestant i's behavior, contestant j's optimal response is to also indirectly lower his effort levels. Because of the symmetry of $g^{nr}\left[\cdot\right]$ around zero, the reduction of the marginal probability of winning is identical for both competitors.

It follows that the direct and indirect effect work in opposite directions on contestant j's incentives, while they work in the same (negative) direction on contestant i's incentives. This means that there must be a certain asymmetry threshold below which competitor j will increase his effort levels and above which contestant j will decrease his effort levels in equilibrium (compared to the symmetric game). This threshold is given by:[14]

$$1 + \alpha = \frac{g\left[0\right]}{g^{nr}\left[\cdot\right]} \tag{4.21}$$

Now, the asymmetry level at which the direct effect outweighs the indirect effect the most, i.e. at which contestant j's equilibrium effort levels are

---

[13] Remember that both probabilities have to add up to unity.

[14] If $1 + \alpha > \frac{g\left[0\right]}{g^{nr}\left[\cdot\right]}$, the direct effect outweighs the indirect effect, so that contestant j increases its effort levels, and if $1 + \alpha < \frac{g\left[0\right]}{g^{nr}\left[\cdot\right]}$, the indirect effect outweighs the direct effect, so that contestant j decreases its effort levels in equilibrium compared to the symmetric game.

maximized, is then given by

$$1 + \alpha = -\frac{g^{nr}[\cdot]}{\frac{\partial g^{nr}[\cdot]}{\partial \alpha}}, \tag{4.22}$$

where $\frac{\partial g^{nr}[\cdot]}{\partial \alpha} < 0$.[15] Conversely, competitor i's equilibrium effort levels will unambiguously decrease because it must always be the case that $(1 - \alpha) <$ $\frac{g[0]}{g^{nr}[\cdot]}$ for $0 < \alpha < 1$, as $g^{nr}[\cdot] < g[0]$.

Regarding the average game quality, however, notice that, due to the symmetry of the density function $g^{nr}[\cdot]$ around zero, the indirect marginal probability effect cancels out and thereby has no effect on the average quality of a game. In other words, because contestants can not only increase their winning probability through productive playing effort but also through sabotage effort, the negative indirect effect of an asymmetry between players on the quality of a contest is nullified. Thus, the change in the game quality, as a result of an asymmetry between competitors, is only driven by the direct effects:

$$\phi_{nr}^a = \frac{\frac{t^\mu(1-\alpha)k}{t^s(1+\alpha)c} + \frac{t^\mu(1+\alpha)k}{t^s(1-\alpha)c}}{2} = \frac{t^\mu k}{t^s c}\frac{(1+\alpha^2)}{(1-\alpha^2)} \tag{4.23}$$

The superscript a in equation (4.23) signifies that this is the quality of an *asymmetric* game. Note that for $\alpha = 0$, equation (4.23) converts back to equation (4.9).

Expression (4.23) demonstrates that the asymmetry in prize valuations in fact increases the quality of a game compared to a symmetric game for all values of α, as $\frac{(1+\alpha^2)}{(1-\alpha^2)} > 1$ for all $0 < \alpha < 1$. This result is explained by the fact that sabotage effort decreases the quality of a contest at a decreasing rate, while productive playing effort increases the game quality at a constant rate. Consequently, the direct effect that decreases contestant i's sabotage

---

[15] For a detailed derivation of the asymmetry level maximizing contestant j's effort levels, please see Appendix A.1.

effort has a greater effect on the game quality than the direct effect that increases contestant j's sabotage effort. At the same time, the equal, yet opposing, direct effects on the contestants' productive playing efforts cancel each other out.

Asymmetric Productive Playing Talents   Consider now another likely asymmetry between competitors, namely an asymmetry in productive playing talent. Similar to the analysis above, suppose that $t_i^\mu = (1 - \beta)t^\mu$ and $t_j^\mu = (1 + \beta)t^\mu$, where $0 < \beta < 1$ represents the asymmetry in productive playing talent. Again, it will be assumed that players are still symmetric in all other characteristics so that the effect of this asymmetry on the quality of a game can be studied in isolation. Given these assumptions, equations (4.7) and (4.8) convert to

$$\mu_{i_{nr}}^* = \frac{(1 - \beta)t^\mu g^{nr}[\cdot]V}{c} \qquad (4.24)$$

$$s_{ij_{nr}}^* = \frac{t^s g^{nr}[\cdot]V}{k} \qquad (4.25)$$

for competitor i's effort levels and to

$$\mu_{j_{nr}}^* = \frac{(1 + \beta)t^\mu g^{nr}[\cdot]V}{c} \qquad (4.26)$$

$$s_{ji_{nr}}^* = \frac{t^s g^{nr}[\cdot]V}{k} \qquad (4.27)$$

for competitor j's effort levels. The optimal strategies (4.24) to (4.27) again illustrate a *direct* and an *indirect* effect.

On the one hand, the increased effectiveness of contestant j's playing effort *directly* increases contestant j's productive effort incentives, while the lower effectiveness of contestant i's productive effort lowers contestant i's productive effort incentives. This leads to a higher equilibrium probability of success for contestant j and an equivalently lower probability of success for contestant i. Note that the sabotage activities by both contestants remain unaffected by the direct effect.

On the other hand, due to the *indirect* effect resulting from the contestants' lower marginal probabilities of success, the asymmetry in productive playing talent at the same time reduces the incentives of both competitors to exert productive and sabotage effort.

Thus, as long as the direct effect outweighs the indirect effect, i.e. as long as $1 + \beta > \frac{g[0]}{g^{nr}[\cdot]}$, the asymmetry in productive playing ability causes contestant j to increase his productive effort relative to a symmetric game. As in the case of asymmetric prize valuations, contestant j's productive effort is maximized at the asymmetry level where $1 + \beta = -\frac{g^{nr}[\cdot]}{\frac{\partial g^{nr}[\cdot]}{\partial \beta}}$, where $\frac{\partial g^{nr}[\cdot]}{\partial \beta} < 0$. Since the direct effect and the indirect effect work in the same (negative) direction for contestant i, his productive effort is unambiguously reduced.

Now, consider the average game quality with an asymmetry in productive playing talent:

$$\phi_{nr}^{a} = \frac{\frac{(1-\beta)t^{\mu}k}{t^{s}c} + \frac{(1+\beta)t^{\mu}k}{t^{s}c}}{2} = \frac{t^{\mu}k}{t^{s}c} \quad (4.28)$$

Comparing expression (4.28) with (4.9), it is readily observed that the quality of the game remains unaffected by an asymmetry in productive playing abilities. Not only the indirect but also the direct effect cancels out in the computation of the game quality.[16]

As expression (4.9) already showed, a game with overall greater productive playing talent has a higher quality than a game with overall lower pro-

---

[16] Of course, this result requires, as is implicitly assumed here, that the total amount of productive playing talent in the game remains unchanged. Note that $(1 - \beta) t^{\mu} + (1 + \beta) t^{\mu} = t_{i}^{\mu} + t_{j}^{\mu}$, which is the same total amount of talent as in the symmetric game studied in Section 4.2.

ductive playing talent. Expression (4.28) now, however, in addition demonstrates that the mere fact that contestants have unequal playing abilities has no effect on the quality of the contest. In other words, the quality of a game, as it is defined in this model, between a strong and a weak contestant will be equal to the quality of a game between two competitors of medium strength, as long as the absolute amount of ability in both types of games is identical.

Asymmetric Sabotage Talents   Let us now examine what happens if the contestants are asymmetric in sabotage talents. One could think of sabotage talent as the talent of hiding sabotage activities from the referee or as the ability of misleading the referee through blatant diving, for instance. Similar to the asymmetry analyses above, assume that $t_i^s = (1 - \gamma)t^s$ and $t_j^s = (1 + \gamma)t^s$ with $0 < \gamma < 1$, while all other asymmetry parameters are kept symmetric. Such an asymmetry in sabotage talent transforms equations (4.7) and (4.8) to

$$\mu_{i_{nr}}^* = \frac{t^\mu g^{nr} \left[\cdot\right] V}{c} \tag{4.29}$$

$$s_{i_{j_{nr}}}^* = \frac{(1 - \gamma)t^s g^{nr} \left[\cdot\right] V}{k} \tag{4.30}$$

for contestant i and to

$$\mu_{j_{nr}}^* = \frac{t^\mu g^{nr} \left[\cdot\right] V}{c} \tag{4.31}$$

$$s_{j_{i_{nr}}}^* = \frac{(1 + \gamma)t^s g^{nr} \left[\cdot\right] V}{k} \tag{4.32}$$

for contestant j. Thus, similar results for the contestants' incentives as in the case of asymmetric productive playing talents are obtained, except that now the direct effect appears in the sabotage effort levels and not in the productive effort levels.

Yet, this leads to the important difference between an asymmetry in sabotage talent versus an asymmetry in productive playing talent, namely their effects on the game quality. Remember that, in the case of an asymmetry in productive playing talent, the positive direct effect on contestant j's productive effort cancelled out with the negative direct effect on contestant i's productive effort because productive playing effort increases the average quality of a game linearly.

This, however, is not the case with an asymmetry in sabotage talent. Although the direct effect increases the sabotage activity of competitor j by the same amount as it decreases the sabotage activity of competitor i, the two direct effects affect the average game quality unequally. In fact, because sabotage effort lowers the quality of a contest at a decreasing rate, the direct effect lowering contestant i's sabotage has a greater positive effect on the quality of a match than the direct effect increasing contestant j's sabotage. In other words, the direct effect on contestant i's sabotage effort marginally increases the quality of the game by more than the direct effect on contestant j's sabotage activity marginally decreases the quality of a match. This is demonstrated by the expression

$$\phi_{nr}^a = \frac{\frac{t^\mu k}{(1+\gamma)t^s c} + \frac{t^\mu k}{(1-\gamma)t^s c}}{2} = \frac{t^\mu k}{t^s c}\frac{1}{1-\gamma^2}, \qquad (4.33)$$

where $\frac{\partial \phi_{nr}^a}{\partial \gamma} > 0$. Therefore, other than with an asymmetry in productive playing ability, the quality of a game actually increases with an asymmetry in sabotage ability for all $0 < \gamma < 1$.

Asymmetric Marginal Costs of Effort   Allowing for asymmetries in marginal costs of effort will effectively lead to the same results as shown in the study of asymmetric talents. This is because one could also interpret a more talented competitor as a competitor with a lower marginal cost of effort, i.e. $t_i^\mu = \frac{1}{c_i}$ and $t_i^s = \frac{1}{k_i}$. Thus, asymmetric marginal costs of effort will have similar effects on the quality of a game as those observed with asymmetric abilities.

4.3.2 An Asymmetric Game with an Honest Referee

So far, we have only looked at asymmetries in a game without a referee. Now, we will study asymmetries in a game with an honest referee in order to examine the effect of asymmetries on the value of a referee.

Consider first the asymmetry in prize valuations again, i.e. $0 < \alpha < 1$. Similar to the analysis of asymmetries in a game without a referee, players are assumed to remain symmetric in all other characteristic parameters, i.e. $t_i^\mu = t_j^\mu = t^\mu$, $t_i^s = t_j^s = t^s$, $c_i = c_j = c$ and $k_i = k_j = k$. This turns equations (4.10) and (4.11) into

$$\mu_{i_{hr}}^* = \frac{t^\mu \left[1 - m(1 + F)\right] g^{hr} \left[\cdot\right] (1 - \alpha)V}{c} \tag{4.34}$$

$$s_{ij_{hr}}^* = \frac{t^s \left[1 - (1 - \overline{m})F\right] g^{hr} \left[\cdot\right] (1 - \alpha)V}{k} \tag{4.35}$$

for competitor i and into

$$\mu_{j_{hr}}^* = \frac{t^\mu \left[1 - m(1 + F)\right] g^{hr} \left[\cdot\right] (1 + \alpha)V}{c} \tag{4.36}$$

$$s_{ji_{hr}}^* = \frac{t^s \left[1 - (1 - \overline{m})F\right] g^{hr} \left[\cdot\right] (1 + \alpha)V}{k} \tag{4.37}$$

for competitor j.[17] The quality of a game with asymmetric prize valuations
and an honest referee is therefore represented by

$$\phi_{hr}^a = \frac{t^\mu k}{t^s c} \frac{[1 - m(1 + F)]}{[1 - (1 - \overline{m})F]} \frac{1 + \alpha^2}{1 - \alpha^2}. \tag{4.38}$$

Hence, using expression (4.23), the value of a referee is given by

$$\frac{\phi_{hr}^a}{\phi_{nr}^a} - 1 = \frac{\frac{t^\mu k}{t^s c} \frac{[1-m(1+F)]}{[1-(1-\overline{m})F]} \frac{1+\alpha^2}{1-\alpha^2}}{\frac{t^\mu k}{t^s c} \frac{1+\alpha^2}{1-\alpha^2}} - 1 = \frac{[1 - m(1 + F)]}{[1 - (1 - \overline{m})F]} - 1. \tag{4.39}$$

Comparing equation (4.39) with (4.16), we discover quickly that the value
of a referee remains unchanged. In other words, an asymmetry in prize
valuations does not affect the value of the referee.

This might seem surprising at first remembering that in expression (4.23)
we observed that the asymmetry in prize valuations increased the quality of
a game. Because the asymmetry in prize valuations in itself already led to
a higher game quality, one might imply that the value of a referee would
be reduced with the extent of the prize valuation asymmetry, as the game
quality is already high even without the referee.

But this mistaken conclusion presumes that the value added by the referee decreases with the quality of a game without a referee. However, this is
not the case. Recall from expression (4.16) that a referee's value is merely
determined by his own performance. Therefore, it does not matter whether
the quality of a game without a referee is already high or low. As long as
an asymmetry between contestants does not affect the referee's errors, the
value of the referee will remain unchanged.

Note that similar to an asymmetry in prize valuations there is no obvious reason to suppose that an asymmetry in the competitors' productive

---

[17]   Remember that $g^{hr}[\cdot] := g[y^{hr}(\mu_i, s_{ji}) - y^{hr}(\mu_j, s_{ij})]$.

playing talents would affect the errors of an honest referee. Consequently, an asymmetry in productive playing talents will have no effect on the value of the referee either.[18]

Conversely, if we want to study the effect of an asymmetry in sabotage talents on the value of the referee, it would be unrealistic to assume that the referee's type 1 and type 2 errors are independent of the contestants' abilities to sabotage. Therefore, a different result will be attained when contestants are asymmetric in sabotage talent. The type 1 errors will increase with the contestants' abilities of misleading the referee (e.g. through blatant diving), while the type 2 errors will increase with their abilities of hiding sabotage activities.

Thus, assume for the sake of the argument now that $\frac{\partial m}{\partial t^s} > 0$ and $\frac{\partial^2 m}{\partial^2 t^s} < 0$ as well as $\frac{\partial \overline{m}}{\partial t^s} > 0$ and $\frac{\partial^2 \overline{m}}{\partial^2 t^s} < 0$. Of course, the referee's errors to the detriment of contestant i will be a function of the sabotage talent of contestant j and vice versa. It follows that, if contestants are asymmetric in sabotage talents, the referee's errors will not be the same for both competitors any more, even if the referee is honest. Therefore, let $t_i^s = (1 - \gamma)t^s$ and $t_j^s = (1 + \gamma)t^s$ so that $m_i(t_j^s) > m_j(t_i^s)$ and $\overline{m}_i(t_j^s) > \overline{m}_j(t_i^s)$. Thus, equations (4.10) and (4.11) transform to

$$\mu_{i_{hr}}^* = \frac{t^\mu \left[1 - m_i(t_j^s)(1 + F) \right] g^{hr} \left[\cdot \right] V}{c} \tag{4.40}$$

$$s_{ij_{hr}}^* = \frac{(1 - \gamma)t^s \left[1 - (1 - \overline{m}_j(t_i^s)) F \right] g^{hr} \left[\cdot \right] V}{k} \tag{4.41}$$

---

[18]   The mathematical support of this argument is presented in Appendix A.2.

for contestant i and to

$$\mu^*_{j_{hr}} = \frac{t^\mu \left[1 - m_j(t^s_i)(1 + F)\right] g^{hr}\left[\cdot\right] V}{c} \tag{4.42}$$

$$s^*_{ji_{hr}} = \frac{(1 + \gamma)t^s \left[1 - \left(1 - \overline{m}_i(t^s_j)\right) F\right] g^{hr}\left[\cdot\right] V}{k} \tag{4.43}$$

for contestant j. Since $\frac{\partial m}{\partial t^s} > 0$, $\frac{\partial^2 m}{\partial^2 t^s} < 0$ and $\frac{\partial \overline{m}}{\partial t^s} > 0$, $\frac{\partial^2 \overline{m}}{\partial^2 t^s} < 0$ the referee's errors to the detriment of competitor j will go down (due to the decrease in contestant i's sabotage talent) by a greater amount than the errors to the detriment of competitor i will increase (due to the equivalent increase in contestant j's sabotage talent). In other words, the asymmetry in sabotage talents improves the performance of the referee. As a result, competitor j will increase his productive effort by more and his sabotage activity by less than contestant i will lower them. This leads to an increase in the game quality not only because of the asymmetry effect illustrated in expression (4.33) but also because of the increased added value by the referee.

Surely, this result heavily depends on the assumption that the referee's errors increase at a decreasing rate with the competitors' abilities to sabotage. If it would be assumed instead that the errors increase at a constant rate, i.e. $\frac{\partial m}{\partial t^s} > 0$, $\frac{\partial^2 m}{\partial^2 t^s} = 0$ and $\frac{\partial \overline{m}}{\partial t^s} > 0$, $\frac{\partial^2 \overline{m}}{\partial^2 t^s} = 0$, the value of the referee will remain unchanged. The additional amount of mistakes the referee will make to the detriment of contestant i will be compensated by the fewer mistakes made to the detriment of contestant j. Thus, the overall amount of mistakes is the same as in a symmetric game. Yet, the quality of the game will still increase due to the remaining direct asymmetry effect on the competitors' strategies, as described by equation (4.33).

Conversely, if it was assumed that $\frac{\partial m}{\partial t^s} > 0$, $\frac{\partial^2 m}{\partial^2 t^s} > 0$ and $\frac{\partial \overline{m}}{\partial t^s} > 0$, $\frac{\partial^2 \overline{m}}{\partial^2 t^s} > 0$, the referee's performance and thereby also his value will go down. In this case, the effect of an asymmetry in sabotage abilities will lead to an ambiguous change in the quality of the game. Depending on whether or not the direct asymmetry effect on the contestants' strategies (leading to a higher game quality) outweighs the asymmetry effect on the value of the referee (in this case leading to a lower game quality), the quality of the game will increase, decrease or remain unaltered.

Proposition 4 *The value of a referee is independent of asymmetries between contestants, as long as the asymmetric attribute does not affect the referee's type 1 and type 2 errors.*

4.4 The Threat of a Corrupt Referee

As several corruption scandals involving referees have shown in the past, it is not unusual for referees to be biased in favor of one of the competing contestants.[19] All ethical and moral arguments taken aside, as soon as the referee's expected benefit exceeds the expected cost from corruption, it is only rational for the referee to use his powerful position to collect illicit rents in the form of bribes.[20]

Therefore, simply suppose now that the referee is biased in favor of contestant i. Assume contestant i bribed the referee while contestant j was not interested in doing so because contestant i has a lower productive playing ability than contestant j, i.e. because $t_i^\mu = t^\mu(1 - \beta)$ and $t_j^\mu = t^\mu(1 + \beta)$.[21]

---

[19] The corruption scandal in the Fußball Bundesliga in 2005 involving the DFB referee Robert Hoyzer (see Ahrens (2005)), the National Basketball Assciation (NBA) scandal involving the referee Tim Donaghy in 2008 (see DPA (2008)) or the scandal in the Handball Bundesliga involving the two DHB referees Frank Lemme and Bernd Ullrich in 2009 (see DPA (2009a)), for example. For more examples of international corruption scandals, see Maennig (2002, 2005).

[20] The incentives for referees and also for contestants to collude will be studied in more detail in Chapter 6.

[21] Note that, because contestant i has a lower equilibrium probability of success, as he will exert a lower productive effort in equilibrium than contestant j, he also has a lower equilibrium payoff from participating in the game in an honest way, so that it is realistic to assume that contestant i is more inclined to bribe the referee than contestant j.

Now, let m_j^b and $\overline{m}_j^b$ be the probabilities that the biased referee makes a wrong type 1 and type 2 call respectively against contestant j. Because the referee cannot lower his unconscious errors, the referee bias must lead to an increase in the referee's type 1 and type 2 errors to the detriment of contestant j by a conscious component. Thus, it must be that $m_j^b > m_i = m$ and $\overline{m}_j^b > \overline{m}_i = \overline{m}$, where, for simplicity, let us suppose that $m_j^b - m = \overline{m}_j^b - \overline{m}$. To simplify the notation below, let $g^{br}\left[\cdot\right] := g\left[y^{br}(\mu_i, s_{ji}) - y^{br}(\mu_j, s_{ij})\right]$, where the superscript br indicates the marginal probability of success in a game with a *biased referee*. The referee bias then results in the following equilibrium strategies:[22]

$$\mu_{i_{br}}^* \;=\; \frac{t^\mu\,(1-\beta)\,[1 - m(1+F)]\,g^{br}\left[\cdot\right]V}{c} \tag{4.44}$$

$$s_{ij_{br}}^* \;=\; \frac{t^s\,\left[1 - (1 - \overline{m}_j^b)F\right]g^{br}\left[\cdot\right]V}{k} \tag{4.45}$$

$$\mu_{j_{br}}^* \;=\; \frac{t^\mu\,(1+\beta)\,\left[1 - m_j^b(1+F)\right]g^{br}\left[\cdot\right]V}{c} \tag{4.46}$$

$$s_{ji_{br}}^* \;=\; \frac{t^s\,[1 - (1 - \overline{m})F]\,g^{br}\left[\cdot\right]V}{k} \tag{4.47}$$

Even though we know that, due to the normal distribution of the composed variable $x := \varepsilon_j - \varepsilon_i$ around zero, $g\left[0\right] > g^{br}\left[\cdot\right]$, it is not clear whether $g^{br}\left[\cdot\right] > g^{hr}\left[\cdot\right]$ or whether $g^{br}\left[\cdot\right] < g^{hr}\left[\cdot\right]$, as it depends on the ex-ante asymmetry level β and thereby on the extent by which contestant i can increase his ex-ante equilibrium probability of success by bribing the referee.

---

[22] For simplicity, it is assumed here that contestant i has a certain amount of excess bribe money available, which could be interpreted as a first mover advantage. This assumption will, however, be relaxed in Chapter 6.

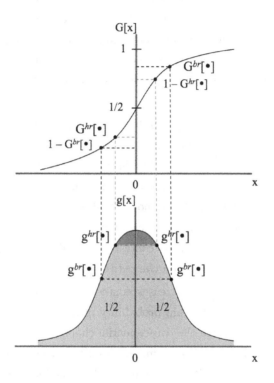

Figure 4.2: The effect of a referee bias on contestant i's and j's absolute and marginal probability of success in a game with a low asymmetry in productive playing abilities

As shown in Figure 4.2, if $G^{br}\left[\cdot\right] > 1 - G^{hr}\left[\cdot\right]$,[23] so that by bribing the referee contestant i can raise his equilibrium chance of winning above the probability of success that contestant j had in a game with an honest referee, $g^{br}\left[\cdot\right] < g^{hr}\left[\cdot\right]$. This would most likely be the case when the asymmetry in productive playing ability β is relatively low. If, however, $G^{br}\left[\cdot\right] < 1 - G^{hr}\left[\cdot\right]$, so that contestant i can only raise his winning chances to a probability of success that is still lower than contestant j's probability of winning in a game with an honest referee, then it would be the case that $g^{br}\left[\cdot\right] > g^{hr}\left[\cdot\right]$. This may occur if β is rather large.

Expressions (4.44) to (4.47) therefore exhibit that the referee bias, on the one hand, leads to an unambiguous *direct* effect, and, on the other hand, to an ambiguous *indirect* effect.[24] The bias directly reduces contestant j's productive effort because the referee more often falsely convicts contestant j's productive effort now leading to more penalties against competitor j. At the same time, the referee also convicts contestant i's sabotage less often, leading to fewer penalties against contestant i and therefore to a higher sabotage effort by competitor i on competitor j.

Because the direct effect increases contestant i's and equivalently decreases contestant j's equilibrium probability of success, the bias *indirectly* affects the effort choices via the consequential change in the marginal probability of success. Because of the symmetry of $g^{br}\left[\cdot\right]$ around zero, the reduction of the marginal probability of winning is the same for both competitors. However, where for competitor i the adjustment of the marginal probability results from an ex-ante advantage, the adjustment of the marginal probability of winning of contestant j results from the ex-ante disadvantage. Thus, if the asymmetry is fairly small so that $g^{br}\left[\cdot\right] < g^{hr}\left[\cdot\right]$, the indirect effect low-

---

[23]   Let $G^{hr}\left[\cdot\right]$ $:=$ $G\left[y^{hr}(\mu_i, s_{ji}) - y^{hr}(\mu_j, s_{ij})\right]$ and $G^{br}\left[\cdot\right]$ $:=$ $G\left[y^{br}(\mu_i, s_{ji}) - y^{br}(\mu_j, s_{ij})\right]$.

[24]   The ambiguity of the indirect effect arises from the uncertainty of whether $g^{br}\left[\cdot\right] < g^{hr}\left[\cdot\right]$ or whether $g^{br}\left[\cdot\right] > g^{hr}\left[\cdot\right]$.

ers the contestants' productive effort and sabotage effort. If the asymmetry is so large that $g^{br}[\cdot] > g^{hr}[\cdot]$, the indirect effect increases the contestants' equilibrium effort strategies.

From this follows that if $g^{br}[\cdot] < g^{hr}[\cdot]$, contestant i's productive effort decreases relative to an asymmetric game with an honest referee, while his sabotage effort increases or decreases depending on whether which of the two opposing effects dominates. While contestant j unambiguously reduces his productive effort and his sabotage effort, he reduces his productive effort by more. This is because both effects work in the same negative direction on contestant j's productive effort, while contestant j's sabotage effort is only indirectly reduced.

Proposition 5 *If the asymmetry between the contestants is relatively small, the biased referee (1) reduces the productive effort of the unfavoured contestant by a significant amount, (2) reduces the unfavoured contestant's sabotage effort and the favoured contestant's productive effort by an equal yet relatively small amount, and (3) increases or slightly decreases the sabotage effort of the favoured contestant.*

If, however, $g^{br}[\cdot] > g^{hr}[\cdot]$, i.e. if the asymmetry in playing ability is so large that even the biased referee can not make sure that $G^{br}[\cdot] > 1 - G^{hr}[\cdot]$, both contestant i's productive effort and his sabotage effort are increased, while contestant i increases his sabotage effort by more (due to the complementing positive direct and indirect effect). While contestant j also increases his sabotage effort by the same amount as contestant i increases his productive effort, contestant j may increase or decrease his productive effort. The direction of contestant j's adjustment in his productive effort depends on which of the two effects dominates, as the indirect effect and the direct effect work in opposite directions.

Proposition 6 *If the asymmetry between the contestants is large, the bi-*

ased referee (1) increases the sabotage effort of the favoured contestant by a significant amount, (2) increases the favoured contestant's productive effort and the unfavoured contestant's sabotage effort by an equal yet relatively small amount, and (3) increases or slightly decreases the productive effort of the unfavoured contestant.

Independent of whether $g^{br}[\cdot] > g^{hr}[\cdot]$ or whether $g^{br}[\cdot] < g^{hr}[\cdot]$, however, the referee bias will unambiguously lower contestant j's overall performance, i.e. $q_j^{hr}\left(\mu_{j_{hr}}^*, s_{ij_{hr}}^*\right) > q_j^{br}\left(\mu_{j_{br}}^*, s_{ij_{br}}^*\right)$. This is because, if $g^{br}[\cdot] < g^{hr}[\cdot]$, both effects work in the same deteriorating direction for contestant j's productive effort while they work in opposite directions for contestant i's sabotage effort. Therefore, even if the indirect effect dominates, the reduction in competitor j's productive effort will be greater than the reduction in competitor i's sabotage effort. Equivalently, if $g^{br}[\cdot] > g^{hr}[\cdot]$, both effects complementarily increase contestant i's sabotage effort on competitor j, while the two effects work in opposite directions for contestant j's productive effort. Thus, even if the indirect effect dominates, contestant i's increase in sabotage effort will be greater than contestant j's increase in productive effort.

Contestant i's overall performance, however, is only affected by the indirect marginal probability effect. Because $g^{br}[\cdot]$ is distributed symmetrically around zero, the indirect effect affects contestant i's productive effort in the same quantitative way, as it affects competitor j's sabotage effort on competitor i. Consequently, contestant i's overall performance will remain unchanged, implying that $q_i^{hr}\left(\mu_{i_{hr}}^*, s_{ji_{hr}}^*\right) = q_i^{br}\left(\mu_{i_{br}}^*, s_{ji_{br}}^*\right)$.[25]

---

[25]   Recall that $q_i^{hr}\left(\mu_{i_{hr}}^*, s_{ji_{hr}}^*\right) = t_i^\mu \mu_{i_{hr}}^*\left[(1 - m_i(1 + F)\right] - t_j^s s_{ji_{hr}}^*\left[1 - (1 - \overline{m}_i)F\right] + \varepsilon_i$, so that the contestants' overall performances $linearly$ increase in their positive playing effort and $linearly$ decrease in the opponent's sabotage effort. Therefore, as $\mu_{j_{hr}}^* - \mu_{j_{br}}^* > s_{ij_{hr}}^* - s_{ij_{br}}^*$, $q_j^{hr}\left(\mu_{j_{hr}}^*, s_{ij_{hr}}^*\right) > q_j^{br}\left(\mu_{j_{br}}^*, s_{ij_{br}}^*\right)$ and, as $\mu_{i_{hr}}^* - \mu_{i_{br}}^* = s_{ji_{hr}}^* - s_{ji_{br}}^*$, $q_i^{hr}\left(\mu_{i_{hr}}^*, s_{ji_{hr}}^*\right) = q_i^{br}\left(\mu_{i_{br}}^*, s_{ji_{br}}^*\right)$.

From this discussion one can already take a firm guess on what impact a biased referee has on the quality of the game. According to expression (4.2), the average game quality with a biased referee is now given by

$$\phi_{br}^a = \frac{1}{2} \frac{t^\mu k}{t^s c} \left[(1 - \beta) \frac{[1 - m(1 + F)]}{[1 - (1 - \overline{m})F]} + (1 + \beta) \frac{[1 - m_j^b(1 + F)]}{[1 - (1 - \overline{m}_j^b)F]} \right]. \quad (4.48)$$

With expression (4.48) it can easily be shown that the quality of a game with a biased referee is lower than the game quality with an honest referee, i.e. that $\phi_{hr}^a > \phi_{br}^a$.[26]

Because the referee can not lower his exogenously given unconscious error to the detriment of contestant i, the bias forces him instead to increase his error to the detriment of contestant j. Therefore, the referee's performance worsens with the extent of the bias. Now, it was already established in expression (4.13) that the game quality decreases, as the performance of the referee deteriorates. Consequently, the game quality with a biased referee must be lower than the game quality with an honest referee.

Proposition 7 *The overall performance of the unfavoured contestant is reduced, while the overall performance of the favoured contestant is unaffected. Thus, a biased referee reduces the average game quality.*[27]

Without going into further detail, this result also suggests that, if the value of an honest referee is already low for a sports association, meaning that a referee's unconscious type 1 and type 2 error are fairly large, or if the reduction in the value of a referee as a result of the bias is quite severe, i.e. $m_j^b - m = \overline{m}_j^b - \overline{m}$ is fairly large, the quality of a manipulated game might

---

[26] For a mathematical proof, please see Appendix A.3.
[27] This result in essence coincides with Tirole's (1986, p. 207) conclusion that side contracting (here between the contestant and the referee) lowers the efficiency of a hierarchy and, hence, ought to be prevented.

even turn out to be lower than that of a game without any referee at all, i.e. $\phi_{nr}^a > \phi_{br}^a$.[28]

Thus, despite the potential expediency of referees, it is of eminent importance that sports associations are able to deal with a major drawback of the institution of referees - that is the possibility of corruption. As demonstrated above, sports corruption is a problem to be taken seriously by sports associations because it lowers the demand for sports contests not only by impairing the sports ethos of fairness and respect but also by directly reducing the game quality.[29]

---

[28] Remember that by assumption a higher unconscious type 1 error m and a higher type 2 error $\overline{m}$ also lead to a higher type 1 error m_j^b and a higher type 2 error $\overline{m}_j^b$ of a biased referee.

[29] Zaugg (2010) further elaborates on the dangers sports corruption can pose from a social perspective.

Chapter 5

The German Criminal Law on Sports Corruption

The question of how German state law currently does and/or should handle sports corruption is a highly controversial and complex topic among legal experts. With the seemingly increasing amount of corruption scandals being discovered in professional German sports leagues, legal scholars have drawn more and more attention to this question.

The following short overview intends to inform the reader about some of the judicial problems that German professional sports associations presently face in their attempt to deter sports corruption. Yet, due to the wide ranging complexity of this topic, this short review certainly can not pay tribute to all the legal issues relating to sports corruption in detail. The discussion will therefore focus on how German criminal law handles sports corruption in particular with respect to the bribery of referees. The reader should, however, be aware that depending on the specificities of an investigated case, corruption cases may also expel elements of offense under German civil law.

However, because these can differ significantly across cases, the prosecution of sports corruption under German civil law is beyond the scope of this dissertation.[1]

The complexity of the legal aspects of sports corruption mainly results from the fact that, up until now, the sports ethos does not constitute a legally protected interest. Thus, there are currently no existing specific regulations on match manipulations such as *"sports fraud"* and bribery in sports under German criminal law.[2] Therefore, one has to take recourse to the general penal regulations.

However, as it turns out, the present general penal regulations of the German criminal law do not provide sufficient protection regarding the bribery of referees in sports contests. True, the payment of money (or other gifts) to referees in order to make them arrange for a particular match result is colloquially called *"bribe"* or *"fraud"*. But in the context of sports contests this fact is legally not within the scope of protection granted by the present penal regulations.[3]

5.1   Corruptibility and Corruption in Business Intercourse (Section 299 Criminal Code)

"Corruptibility" and *"corruption"* are offenses which always involve two sides, where both sides are offenders. Thus, the relevant facts and circumstances comprise both corruptibility and corruption.

In Germany, Section 299(1) Criminal Code regulates corruptibility. The prerequisite for an offense is that *"(...) an employee or agent of a business,*

1   For a discussion on the civil liability of sports referees, see Heermann (2009).
2   Wabnitz and Janovsky (2007, Marg. No. 137); for a discussion on the pros and cons of developing a legally protected interest in sports, see König (2010) and Kudlich (2010).
3   Klimke (2005), Schlösser (2005, p. 424), Wabnitz and Janovsky (2007, Marg. No. 137), Zieher (2009, pp. 30 et seq.)

demands, allows himself to be promised or accepts a benefit for himself or another in a business transaction as consideration for according an unfair preference to another in the competitive purchase of goods or commercial services (...)".

Now, it may be left open whether and to what extent organized professional and amateur sports, despite the sometimes considerable economic interests involved, must be viewed as a *"business"*. In any case, there is no competitive situation in the *"purchase of goods or commercial services"* in the sense of Section 299(1) Criminal Code, because the result of a match neither constitutes a *"good"* nor a *"commercial service"* within the meaning of Section 299(1) Criminal Code. Besides, the result of a match can not be *"purchased"*, at least not legitimately.[4]

Section 299(2) Criminal Code regulates corruption. It basically requires the same prerequisites as Section 299(1) Criminal Code, except that this regulation penalizes the corruption (and not the corruptibility) of employees or agents in business relationships. Hence, due to the above mentioned restrictions of the prerequisites for an offense to Section 299 Criminal Code, the elements of this offense do not comprise the corruption of referees by a club manager, coach or player either.

Thus, the corruption of referees and their corruptibility in sports competitions can neither be subsumed under Section 299(1) nor under Section 299(2) Criminal Code.

5.2 Criminal Offenses in Office (Sections 331 et seq. Criminal Code)

Sections 331 et seq. Criminal Code penalize the corruption and the corruptibility of persons in public office. The applicability of these regulations can

---

[4]   Schlösser (2005, p. 424), Zieher (2009, p. 30)

just yet be excluded because referees of sports contests, no matter whether they supervise professional or amateur games, are neither officers nor persons specially obligated for the civil service, so that they are no "*arbitrators*" in the sense of Sections 331 et seq. Criminal Code.

In particular, sports referees are neither state judges, nor are they concerned with supervising or deciding legal cases in arbitration proceedings according to Sections 1025 et seq. Code of Civil Procedure (ZPO). Accordingly, punishability according to Section 339 Criminal Code (perversion of justice) does not apply either.[5]

Thus, Sections 331 et seq. Criminal Code also do not provide any elements of offense for the bribery of referees in sports competitions.

5.3   Fraud (Section 263 Criminal Code)

Stipulating the corruption and corruptibility of sports referees as fraud in the sense of Section 263 Criminal Code is also doubtful.

5.3.1   The Referee as the Principal Offender

It is the central task of a referee in professional or amateur sports to supervise the sports competitions in accordance with the rules. Thus, one might hold the opinion that the "*deceit*" committed by a referee who willfully makes an irregular decision, which is required to fulfill the elements of fraud, is constituted by the fact that he pretends his willfully wrong decision to be regular and thus disregards his duty to enforce the rules of the game.[6]

However, the questions whether this must already be considered an act of deceit in the sense of Section 263 Criminal Code and who has been deceived may be left open, since the unanimous view is that such an act lacks a "*disposition of property*" by mistake and a "*property loss*" directly resulting

---

[5]   Zieher (2009, p. 29), Schlösser (2005, pp. 423 et seq.)
[6]   Zieher (2009, p. 32)

therefrom.[7]

Moreover, the bribery of sports referees generally lacks the required *"identity of substance"* between the referee's advantage and the (presumed) disadvantages of possibly deceived persons.[8] *"Identity of substance"* in the sense of Section 263 Criminal Code means that the offender (here: the referee) strives for a pecuniary benefit immediately transferred from the property of the injured party.[9] This means that the offender's benefit directly arises from the expense of the injured property (e.g. the property of the disadvantaged contestant). But this is just what is not the case when a referee is bribed. The referee accepts money from a third party (i.e. contestant i) and by his irregular behavior inflicts damage on another party (i.e. contestant j) or its property. In such cases there is no identity of substance, since the offender acts with the intention of being rewarded for his deceit by a third party.[10]

Thus, if a referee interferes with a match willfully and irregularly because he is granted a pecuniary benefit for doing so, he does not yet automatically commit fraud.

5.3.2   The Referee as a Secondary Offender

In the past, however, referees have already been punished for *"aiding and abetting fraud"* as secondary offenders based on Section 27 Criminal Code. Such cases normally dealt with so-called *"betting fraud"*. The previously mentioned *"Hoyzer"* case, which enjoyed a wide media coverage, is probably the most famous case in Germany lately.[11]

---

[7]   For more information on the problem of immediacy in case of fraud, see Jäger (2010, pp. 761 et seq.).

[8]   Zieher (2009, p. 33)

[9]   Schönke and Schröder (2010, Section 263, Marg. No. 168)

[10] Schönke and Schröder (2010, Section 263, Marg. No. 168)

[11] For more details on the Hoyzer case and the Supreme Court's judgment, see BGH (2007, pp. 782 et seq.).

In this case the principal offender had bet on several soccer matches he had manipulated by paying considerable amounts of money to the also accused referees Hoyzer and Marks, thereby making them fix the results of the soccer games by deliberately irregular calls so that the accused could pocket high winnings. The Supreme Court (BGH) regarded the match manipulation as fraud to the disadvantage of the betting providers concerned in this case. The Supreme Court acted on the assumption that, when a sports bet is made, each of the parties involved tacitly declares that the risk inherent in the bet will not be changed in his favor by any manipulation of the sports event which he has induced and of which the contractual partner has no knowledge.[12]

Thus, the principal offender's deception of the employees of the betting agency concerned constituted deceit implied by his activity.[13] Due to this implied deceit that the object of the bet (i.e. the soccer match) was not manipulated, the employees of the betting agency concerned were under a pertinent misapprehension. The misapprehension of the employees of the betting agency that the matches were not manipulated led to a pecuniary disposition to the betting provider's disadvantage, i.e. to accepting the bet.[14]

This pecuniary disposition caused by deceit also inflicted damage on the betting provider. The bets were sports bets with fixed odds (so-called "*oddset*" bets), in which the quota that had been assessed by the betting provider on the basis of a particular risk constituted, so to speak, the "*sales price*" of the betting odds. The betting odds determine by which factor the stake money is modified if the bet is won. If someone fixes a match, the betting risk is quite substantially shifted to his advantage, since the odds fixed by the betting provider upon the conclusion of the contract do not correspond any

---

[12]   BGH (2007, p. 784) with reference to BGHSt 29, 165 = NJW 1980, 793; for critical views, see Krack (2007, pp. 103 et seq.).

[13]   BGH (2007, p. 785)

[14]   BGH (2007, p. 785)

more to the risk on which the betting provider had based his own commercial calculations.[15]

Whenever a betting contract is concluded, such a "*difference in odds*" constitutes quite a considerable pecuniary damage.[16] Thus, the principal offender in the Hoyzer case was convicted of fraud according to Section 263 Criminal Code, and the referees involved were sentenced for aiding and abetting fraud.

A consequence from this case was that the Federal Conference of the DFB, in an extraordinary meeting on April 28, 2005, determined modifications of its rules and articles in order to ensure the integrity of sports competitions in the future. These reformative resolutions introduced a prohibition of betting for players, officials and referees.[17] Moreover, Section 6a Rules of Law and Procedure (RuVO) now explicitly brands match manipulations by referees, players, coaches and other functionaries as unfair conduct.[18]

This stipulation, however, covers only betting fraud both under criminal law and under DFB sports law, whereas the corruptibility of referees connected with the manipulation of competitions in championships, relegations and so on continues to go unpunished.

5.3.3 The Competitor as the Principal Offender

It is also debatable whether Section 263 Criminal Code results in the punishment of at least club managers, coaches or players for fraud who bribe a referee to manipulate a match.

Prosecution of the club itself under criminal law can be excluded because

---

[15] BGH (2007, p. 785)
[16] BGH (2007, p. 785); for critical views, see Saliger, Rönnau and Kirch-Heim (2007, pp. 361 et seq.).
[17] Section 1(2) Rules of Law and Procedure (RuVO); see also Zieher (2009, p. 27).
[18] Zieher (2009, p. 31)

only natural persons can be punished as offenders.[19] Thus, if at all, it is the functionaries of the club who can be held responsible under criminal law.

The act constituting fraud is deception. As in the Hoyzer case, one might claim that there is an implied deception because in the scope of a competition a club manager or player tacitly declares that he has not bribed the referees assigned to the match for the benefit of his club or himself.

The question is, however, whether the fact that a club manager/player has bribed a referee solely to manipulate a game already causes a direct pecuniary damage in the sense of Section 263 Criminal Code. This depends on whether bribing a referee solely to manipulate a match already results in an endangerment of property that amounts to a damage.

In its Hoyzer decision, the Supreme Court explicitly pointed out that the mere manipulation of a betting game does not yet imply a concrete endangerment of property, which under economic aspects already constitutes an impairment of the present property situation. According to the court, such a concrete endangerment of property amounting to a damage under criminal law is only given if the deceived party seriously has to expect economic disadvantages.[20]

This, according to the court, again must be denied if the occurrence of those economic disadvantages is not even predominantly probable but depends on future events which, despite the manipulation, are still beyond the influence of the bribed party/defrauder to a quite essential extent.[21]

In practice, manipulation attempts fail relatively often. Thus, despite the manipulation, success is not predominantly probable, since it depends on further future events beyond the offender's control. Because of the random factors represented by the random variables ε_i and ε_j in the economic model discussed in Chapter 4, the manipulation of a match therefore merely

---

[19]   Schönke and Schröder (2010, ante Sections 25 et seq., Marg. No. 119)
[20]   See BGH (2007, p. 786) for more information.
[21]   BGH (2007, p. 786)

constitutes an abstract endangerment of property.[22] An abstract endangerment of property, however, is not sufficient for assuming a pecuniary damage as required by the definition of fraud.[23]

Thus, the Supreme Court takes up the position that bribing a referee does not yet result in an impairment, under economic aspects, of the present property situation of the opposing contestant. Thus, the corruption of referees by club managers, coaches or players cannot be regarded as fraud because a concrete endangerment of property that amounts to a damage is missing. As long as there is no experience yet on what conditions a successful manipulation by a referee depends, this point of view will probably not change.[24]

It follows that, even though the characteristics of each individual case (e.g. in case of betting fraud) may, in exceptional cases, result in the application of the definition of fraud, there is no general applicability of Section 263 Criminal Code in the scope of manipulations of sports contests.[25]

5.4 Embezzlement (Section 266 Criminal Code)

Finally, the offense of embezzlement according to Section 266 Criminal Code might be considered.

5.4.1 Club Managers as Offenders

If a club manager bribes a referee, he commits an act of embezzlement, because due to his action the property of the club is misused. A club manager has to safeguard the pecuniary interests of the club. When bribing a referee, he violates this duty, since his disposition of the property of the club is con-

---

[22] BGH (2007, p. 786)

[23] This judicial view is based on the general and fundamental principle "*nulla poena sine lege*" ("*no punishment without law*") underlying the regulations of the Criminal Code. It prescribes present regulations to be interpreted only in a narrow sense, disallowing attempts of extensive analog application.

[24] Krack (2007, pp. 111 et seq.)

[25] A detailed presentation of the problems can be taken from Duyar (2009).

trary to its rules and articles.[26] In 2006, for instance, the former managing director of Bayer04 Leverkusen, Reiner Calmund, was investigated for embezzlement, because he had aroused suspicion to have used €580,000.- of his club to influence match results in the Fussball Bundesliga.[27]

However, the problematic aspect in such cases is always the pecuniary disadvantage required for a conviction according to Section 266 Criminal Code, because a "damage" may not apply if the club, due to its manager's unlawful act, concurrently receives an equivalent benefit balancing the pecuniary loss or even exceeding the economic amount of the loss.[28]

On February 27, 1975, the Supreme Court decided on a case in the Fussball Bundesliga, in which the president of the club Arminia Bielefeld, with the approval of some board members, had withdrawn DM100,000.- from the club account. This amount was used to bribe players of the opposing club Hertha BSC Berlin.[29] For Arminia Bielefeld it was a matter of staying in the first league, which was and still is of substantial economic importance.

As in case of fraud, the definition of embezzlement requires to assess the inflicted disadvantage in the scope of a damage survey by comparing the property the party attended to would have without the offender's breach of duty with the property the party attended to has as a result of the breach of duty. This comparison has to take into account any benefit gained by the disloyal act. Under the relevant economic aspects, "property" comprises anything that can be measured in monetary value.[30]

Thus, the only decisive criterion in the assessment of a pecuniary disadvantage is the question of whether the fact that the club stays in the first league, which has been achieved by manipulation, makes up for the sacrificed

---

[26]   BGH (1975, p. 1234), Triffterer (1975, p. 613)
[27]   Wabnitz and Janovsky (2007, Marg. No. 136)
[28]   Triffterer (1975, pp. 613 et seq.), Schreiber and Beulke (1977, pp. 658 et seq.)
[29]   For more information on the facts of the case, see BGH (1975, pp. 1234 et seq.).
[30]   BGH (1975, p. 1235)

property of the club (i.e. the bribes and the risk of revocation of the license due to the manipulation). A pecuniary damage would only have to be assumed if the offender, like a gambler, took an extreme risk of loss, willfully and contrary to the rules of mercantile diligence, just to gain a prospect of highly doubtable winnings.[31]

According to the Supreme Court, this assessment finally has to evaluate whether the impending detection of the manipulation and the thus more or less definite expectation of losing the license is more probable than the expectation of accruing profit.[32] Thus, the Supreme Court is exclusively guided by the economic aspect of embezzlement as a crime against property. If there is no pecuniary disadvantage because the accrued profit to be expected from the manipulation is higher than the risk of loss if the license is revoked, no embezzlement has been committed.

According to current criminal law, a match manipulation therefore has to be regarded as a speculative enterprise. Each time when, under economic aspects, the probability of a profit exceeds that of a loss in the form of revocation of license, the manipulation goes unpunished. This means that, when spending monies of the club for manipulation purposes, a club manager must thoroughly consider whether the manipulation is indeed worth its costs.

5.4.2 Referees as Offenders

Contrary to a club manager, a referee is not subject to any special duty to take care of property. He is only obligated to correctly supervise the match, but not to safeguard the pecuniary interests of the persons (including spectators), clubs or federations that are in any way involved in the match.[33]

Therefore, if a referee interferes with a match willfully and irregularly

---

[31] BGH (1975, p. 1236)
[32] BGH (1975, p. 1236)
[33] Zieher (2009, p. 33), Von Komorowski and Bredemeier (2005), Kuhn (2001, p. 109), Heermann (2009, pp. 56 et seq.)

because he is granted a financial advantage, he does not commit embezzlement in the sense of Section 266 Criminal Code due to lack of duty to take care of property.

5.5   Cooperative Prevention of Corruption by State Power and Federation Authority

5.5.1   Autonomy of Associations

Although German sports federations, in particular after the betting scandal of 2006, made stronger endeavors to detect manipulation attempts in competitions by monitoring systems and to prevent manipulations, to date one has not succeeded in containing manipulations of sports competitions in a really effective way. The consequence is that the calls for the definition of a criminal offense of "*sports fraud*" are growing ever louder.[34]

On the basis of the autonomy of associations guaranteed by Section 25 Civil Code (BGB) and Art. 9 para. 1 Basic Constitutional Law (GG), sports clubs and federations already have some regulatory authority and punitive power at their command. "*Autonomy of associations*" means that a club/federation can itself regulate its own organization and the legal relationship between itself and its members in the scope of legal provisions, within the general limits of private autonomy and the general principles of corporate law in a way binding all members. In this sense, autonomy of associations includes the right of an association to its own legislation, above all by articles of association and subsidiary regulations, as well as the right of an association to self-administration by applying and enforcing its own legislation in individual cases.[35]

Thus, the autonomy of associations entitles clubs and federations to stip-

---

[34]   cf. König (2010), Wegmann (2010), or Süddeutsche Zeitung (2006).
[35]   Summerer (1998, p. 80)

ulate rules of conduct and to enforce them by means of its own regulatory authority and punitive power.[36]

To enforce their rules, clubs/federations avail themselves of a variety of sanction instruments, e.g. warning, reprimand, fine, banning from the playing field, removal from office, exclusion from the club, permanent or temporary suspension, deduction of scores, disqualification, revocation of license, or compulsory relegation.[37] Yet, the above-mentioned possibilities of sanction, as differentiated measures with reference to sports, only seem moderately severe compared to the catalogue of punishments provided by the Criminal Code including fines and imprisonment.

Furthermore, the autonomy of associations limits the sanctionary power of associations to its members.[38] Thus, by voluntarily quitting his membership from a club underlying the authority of the association, a wrongdoer can simply escape the association's sanctions. This is exactly what the accused soccer referee Robert Hoyzer chose to do.[39] However, this is not to say that the association can not again take up the prosecution of previously exited wrongdoers in case of re-entry.

The more important reason why clubs/federations also demand the intervention of the state, however, is that they lack an effective set of instruments for investigating irregularities. Only in the scope of criminal investigation could coercive methods used in criminal procedure, such as search, attachment, remand, wire-tapping etc., be applied.[40] Contrary to the government, clubs/federations do not have any administrative or jurisdictional authority in this respect. Thus, clubs and federations hope that a specification of

---

[36] Summerer (1998, p. 139)
[37] Summerer (1998, p. 139)
[38] cf. DFB (2010a, Section 10)
[39] Handelsblatt (2005), Spiegel (2005); as already mentioned, because the Hoyzer case involved betting fraud, which is regulated by §263 Criminal Code, he was nevertheless prosecuted by the ordinary courts.
[40] Wabnitz and Janovsky (2007, Marg. No. 137), Justiz Bayern (2009, p. 2)

sports corruption as a criminal offense changes the general sense of right and wrong, which is supposed to result in a heightened general prevention.[41]

Antagonists, however, support the view that an intervention of the state is not justified, mainly because they do not consider the sports ethos as a legal interest worth protecting.[42] Instead, they argue that it is the responsibility of self-governed sports to cope with manipulations in sports contests (including doping).[43] In this way, the concretization and protection of special values in sports shall solely be reserved to organized sports.

The problem is, however, that an unlimited autonomy of sports and the transfer of the exclusive legislation authority to sports associations would make sports an area beyond the law. Furthermore, irregularities in sports are already now punished under criminal law, for instance if gross fouls committed in sports competitions are qualified as bodily harm. This is why the final right of control always has to remain reserved to the state.[44]

In addition, supporters of creating a special criminal offense definition against corruption in commercial sports point out that sports in modern society have an eminent position and are thus often sponsored also by the government. They claim that sports make an important contribution to the solidarity of society, such as the integration of migrants, the socialization of children and adolescents as well as health protection. In particular the requirement of fairness in sports has a signaling effect in the scope of the ethical intention of equal opportunities. Thus, they argue that future regulations ought to start right from the violation of the sports ethos.[45] As sports are often sponsored by the government, there is in fact also a particular public interest in the protection of the integrity of sports. If competitions are ma-

---

[41]   Wabnitz and Janovsky (2007, Marg. No. 137)
[42]   Kudlich (2010)
[43]   Turner (1992, p. 123)
[44]   Smirra (2008, p. 2)
[45]   Kargl (2007, pp. 494 et seq.)

nipulated to a broad extent, the legitimacy of such governmental sponsoring is lost.

By defining a specific criminal offense against corruption in commercial sports, federations and the state together could take concerted action against the socially harmful impact of manipulated contests. Sports corruption not only inflicts immense losses on sports associations as the organizers of sports contests (as was demonstrated in Chapter 4) but also on private as well as public sponsoring because it is detrimental to the marketing of sports and impedes professional athletes in the exercise of their profession.[46]

5.5.2 The "Draft Bill"

In view thereof, proposals have been made again and again in the past for the definition of special criminal offenses in sports.[47]

The judiciary of the Free State of Bavaria has prepared an extensive draft of a Federal Sports Act (ministerial draft bill on fighting doping and corruption in sports (as per November 30, 2009), hereinafter referred to as the "Draft Bill") according to which doping, bribery, corruptibility and other fraudulent manipulations in sports are prosecuted under criminal law. Moreover, it allows to exhaust all possibilities provided by criminal procedure, including wire-tapping, when so-called sports defrauders are investigated.[48]

According to the Draft Bill, the act is intended to apply to professional sports only because it would be irreconcilable with the prohibition of excessive sentences imposed by the state if the entire mass sports were included as well.[49]

Furthermore, criminal law is now intended to cover any corruption in

---

[46] Wabnitz and Janovsky (2007, Marg. No. 137)
[47] Fritzweiler (1998, pp. 234 et seq.), König (2010, p. 106), Kudlich (2010, p. 108), Wegmann (2010, p. 242), Süddeutsche Zeitung (2006)
[48] Justiz Bayern (2009, pp. 1 et seq.); see also Kudlich (2010, p. 108).
[49] Justiz Bayern (2009, ad Art. 1 para. 1 (Definitionen), p. 18); Wegmann (2010, pp. 244 et seq.) suggests that it should be extended to amateur sports as well.

sports. Art. 1 para. 6 of the Draft Bill defines a special criminal offense for *"corruptibility and corruption in sports"*, which would eliminate the large impunity of corruption in the context of sports contests. Now, for example, briberies of referees in the Fussball Bundesliga that are only aimed at manipulating the contest in the battle for the championship, against relegation, or for the qualification for the Champions League or the UEFA Cup can be punished in the future.

In particular, the Draft Bill proposes to punish such an offense with up to five years of imprisonment or a suitable fine. In severe cases, which according to the Draft Bill arise if an act concerns a particularly extensive pecuniary advantage or if the perpetrator acts commercially or as a member of a group formed particularly for the commitment of such acts, the Draft Bill proposes a punishment of at least six months and up to ten years of imprisonment.

The structure and the wording of the Draft Bill are oriented to the definition of corruptibility and bribery in business intercourse, i.e. of Section 299 Criminal Code.[50]

According to the Draft Bill, organized crimes fall within the terms of so-called extended forfeiture according to Section 73d Criminal Code. This possibility of siphoning off profits gained from criminal activities is intended to fight so-called organized crime more effectively. As substantiated in the Draft Bill, an improved siphoning of profits applies the axe to the mainspring of criminal activities.[51]

In particular cases of serious crimes, the Draft Bill even provides for the possibility of wire-tapping according to Section 100a Code of Criminal Procedure (StPO).[52]

Hence, although some scholars oppose a legally protected interest in

---

[50]   Justiz Bayern (2009, ad Art. 1 para. 6, p. 24)
[51]   Justiz Bayern (2009, Art 1. para. 7(1), p. 8)
[52]   Justiz Bayern (2009, Art. 2, p. 9); see also Kudlich (2010, p. 108).

sports,[53] the Draft Bill would certainly provide prosecutive clarity with respect to sports corruption. The present general penal regulations simply do not provide sufficient protection regarding the bribery of referees in sports contests.

The prospects of success of this Bavarian Draft Bill are, however, difficult to assess. Bavaria already made an advance to punish doping in sports four years ago and failed. Yet, the new Draft Bill exceeds the previous one by far.[54]

---

[53] cf. Kudlich (2010) and Turner (1992)
[54] König (2010, p. 106), Prantl (2009)

Chapter 6

Incentives for Sports Corruption

The apparent loophole in the present regulations provided by state law compels sports associations to themselves deal with the issue of sports corruption, which only arises as a result of an agency problem. Even though a sports association can typically observe and evaluate a referee's objective performance, it can not observe whether a referee made an error deliberately or only unconsciously.[1]

This leaves some room for moral hazard by the referee because the referee can exploit this information asymmetry by entering a side contract for collusion with a contestant.[2] The referee might choose to agree with the contestant to manipulate the game in his favor in return for a share of the

---

[1]   For more information on conflicts with unobservable behaviors (hidden actions) in a principal-agent relationship, see Jost (1999, pp. 286 et seq.).

[2]   This idea was also inferred from Tirole's (1986) and Bac's (1996) study on coalitions in hierarchies of organizations. For more information on opportunistic behaviors in organizations, see also Jost (2000b, pp. 481 et seq.).

winner's prize (i.e. the bribe).

As demonstrated earlier, such collusive side contracting is undesirable for sports associations, as it undeniably lowers the game quality. However, so far the contestants' incentives to bribe a referee were simply assumed to be exogenously given. Likewise, a referee's incentives for corrupt endeavors were not further endogenized. Yet, a sound understanding of both, a contestant's and a referee's incentives for corrupt behavior, will be necessary for sports federations to be able to assess and develop efficient anti-corruption mechanisms.

6.1 Costs and Benefits From Corruption

6.1.1 For the Contestant

As was shown at the end of Chapter 4, the main reason why a contestant would bribe a referee is that, by doing so, he can (substantially) improve his chance of winning the contest. Bribing the referee thereby increases a contestant's expected benefit from competing in a contest.

But, as it is, no benefit comes without cost. A corruptible referee will most likely expect a kickback in return for the referee's bias in the contestant's favour. Yet, because of the many external factors beyond the control of the referee that can still affect the outcome of a game, it would be unrealistic for the contestant to assume that, only because he bribed the referee, he will be ascertained to win the match.[3] For this reason, the contestant would typically condition the payment of the bribe (or at least a substantial fraction of it) on his success in the contest, so that he only transfers the bribe money if he actually wins the contest.

Nevertheless, a contestant attending to bribery will still incur miscel-

---

[3] This uncertainty is exactly the reason why the German Supreme Court viewed the mere manipulation of a game only as an abstract endangerment and not as a concrete endangerment of property in the Hoyzer case.

laneous costs from corruption such as the moral costs incurred from his unsportive behavior, searching costs (i.e. the costs of finding a corruptible referee and actually closing the deal), arising opportunity costs and ultimately, of course, the risk of exposure, which could materialize into a severe penalty.[4]

6.1.2   For the Referee

If the referee indeed agrees to a side contract offered by a contestant, he can collect an illicit rent on top of the standard remuneration he is being paid by the sports association.

However, a referee also incurs miscellaneous costs from corruption such as moral costs and opportunity costs, although these do not need to be identical to those incurred by the contestant. Additionally, a referee has to take the consequences of his present behavior on his ability to pursue his profession in the future into account. In particular, if he enters a side contract, he runs the risk of not being assigned to another game in the future.

In this context, it is noteworthy that according to the rules and articles of most sports federations in Germany, professional sports games can not be supervised by just any referee.[5] Similar to athletes, a referee must have proven his proficient skills over a long period of time in a competitive environment to reach the "*professional*" level.

In fact, there are many individuals who are currently refereeing games at the amateur level, but who would probably like to be able to referee professional games as well. In addition to the high level of prestige awarded by the assignment to professional games, the remuneration for supervising a professional game is much higher than that for amateur games. Nevertheless,

---

[4]   However, remember from the legal discussion that currently the risk of being caught tends to be fairly low because sports federations lack efficient instruments of investigation.

[5]   cf. DFB (2010a, Section 4), DBB (2002, Section 2), or DHB (2007, Section 4)

similar to amateur referees, most sports associations in Germany still do not hire professional referees on a long term basis. Instead, they typically pay their referees a fixed fee per game supervised.[6]

Via careful inspection and selection, a sports association usually holds a list of a limited number of referees who, according to its own standards, are talented and qualified enough to supervise professional games.[7] Once a member of this relatively small number of privileged referees, a referee has to undergo a lot of further skill enhancing training sessions in order to keep up the high quality standard.[8] However, the mere attendance to these training sessions is no guarantee for being assigned to professional games in the future. Instead, the number of professional games a referee gets to supervise by large depends on his performance relative to other listed referees.[9]

Because a corrupt referee has to raise his error rate (above his unconscious error rate) in order to be biased in favor of a contestant, he impedes the exercise of his own profession as a result of a lower probability of being assigned to another game in the future.

6.2 The Model

Having identified a contestant's and a referee's major costs and benefits from side contracting, we could now simply go on and come up with various anti-corruption policies that lower and ideally annihilate the contestant's and/or the referee's incentives for corruption. From an economic perspective corruption deterrence is in the end only a matter of raising the costs of corruption above its benefits.[10]

Yet, merely relying on such a simple cost-benefit analysis is, at least

---

[6]   cf. Koch and Maennig (2007, p. 55), DBB (2008, Section 12), or DHB (2009b, p. 8)
[7]   cf. DFB (2010b, Section 51(4)), or DBB (2002, Section 8)
[8]   cf. DFB (2010a, Section 7(2)), DBB (2002, Section 9), or DHB (2007, Section 3)
[9]   cf. DFB (2010a, Section 4), Koch and Maennig (2007, p. 53)
[10] For such a simplified analysis, see Maennig (2002, 2005).

from an analytical point of view, unsatisfactory because it fully neglects the interdependencies between the referee's and the contestant's incentives. Such interdependencies are, however, crucial for sports associations to be able to implement cost efficient incentive mechanisms to prevent collusive side contracting. To show this, consider the following more sophisticated framework.

Assume the referee can either elicit a high performance by acting in the interest of the sports association and eliciting only the unconscious error m, or he can display a low performance by entering a side contract and eliciting a higher (conscious) error m^b. For simplicity, it will not be distinguished between a type 1 or a type 2 error in the referee's strategy space here any more, as it is not necessary for the line of argument of this model. Thus, let the referee's strategy space be given by $z \in \{m, m^b\}$, where $m < m^b$.

Although the sports association can not observe whether the referee wilfully makes a wrong call or not, assume it receives an imperfectly correlated signal $v_r \in \{v_r^L, v_r^H\}$, where $v_r^L < v_r^H$, on the subjective performance of the referee. More specifically, let the probability distribution of the signal v_r be $\Pr\left(v_r^L \mid m\right) = p_r$, $\Pr\left(v_r^H \mid m\right) = 1 - p_r$, $\Pr\left(v_r^L \mid m^b\right) = 1$ and $\Pr\left(v_r^H \mid m^b\right) = 0$.

It follows that, if the sports association observes a signal $v_r = v_r^H$, which occurs with probability $1 - p_r$, the sports association can be assured that the referee is honest. However, if the sports association observes a signal $v_r = v_r^L$, it does not know whether this is due to the fact that the referee is biased or whether the referee was despite honesty just having a bad day.

Similar to the referee, suppose the contestant can also only choose a low or a high effort so that his strategy space is given by $q \in \{q^L, q^H\}$, where $q^L < q^H$. For simplicity and without loss of generality, it is not distinguished between productive and sabotage effort here any more.

Before the contestant chooses his effort level, assume that he decides whether or not to bribe the referee with an amount $B(z)$, which the referee either rejects or accepts. Hence, by the time the contestant chooses his effort level, the contestant is informed about the referee's state of bias, which, as we already know from Chapter 4, is an important piece of information the contestant takes into account in his effort choice.

Because the referee can only receive a bribe $B(z)$ from the contestant if he agrees to the side contract, let $B(m) = 0$ and $B(m^b) > 0$. To simplify this model further, it will be assumed that the referee can only enter a side contract in the current period so that he does not worry about the potentially lost illicit future income from corruption arising from the risk of not being re-assigned.

Now, assume the contestant conditions the payment of the bribe on the success of the deal, which, from the perspective of the contestant, only occurs if the contestant wins the game. It should be irrelevant in this regard whether or not the contestant would have won the game even without the deliberate help of the referee. As already pointed out, it is impossible for a third party to tell whether a referee deliberately makes a wrong call or not. Thus, the probability that the contestant pays the bribe to the referee is essentially equivalent to the contestant's probability of winning the contest.

The probability that the contestant wins the game shall be given by $G(q, z)$, where $G(q, m) < G(q, m^b)$ and $G(q^L, z) < G(q^H, z)$. Suppose that, if the contestant bribes the referee, he can increase his probability of success by more than if he displayed a high performance, so that $G(q^H, m) - G(q^L, m) < G(q^L, m^b) - G(q^L, m)$, or equivalently, $G(q^H, m) < G(q^L, m^b)$. Assume as well that the higher the probability of success already is, the harder it will be to further increase it by any means. Thus, if the referee is honest, exerting a high effort as opposed to a low effort allows the contestant to increase his probability of success by more than if the referee was biased, i.e. $G(q^H, m^b) - G(q^L, m^b) < G(q^H, m) - G(q^L, m)$.

In addition, assume that, if the contestant wins the contest, he is awarded a winner's prize V, while the sports association pays the referee a legitimate fee S per game supervised.

To account for the fact that a sports association assigns referees based on their objective performance relative to other listed referees allowed to supervise professional games, suppose that the probability that the referee is assigned to another game in the future is given by $H(v_r)$, where $H\left(v_r^H\right) > H\left(v_r^L\right)$. Thus, let $E\left[H(v_r)\right] = p_r H\left(v_r^L\right) + (1 - p_r) H\left(v_r^H\right)$ be the referee's expected probability of re-assignment if the referee is honest, i.e. if $z = m$, and $E\left[H(v_r)\right] = H\left(v_r^L\right)$ if the referee is corrupt, i.e. if $z = m^b$.[11]

Moreover, both, the contestant and the referee, respectively incur an additional cost k_c and k_r from corruption, where $k_r\left(m^b\right) > 0$, $k_c\left(m^b\right) > 0$ and $k_r(m) = k_c(m) = 0$. Let k_c and k_r (among others) not only include the aforementioned miscellaneous costs, such as possible searching costs, moral costs or opportunity costs arising from corruption, but also the expected penalty in case of conviction.

Besides, the contestant and the referee are assumed to incur effort costs. While the referee incurs the same effort cost c_r no matter whether he is biased or not,[12] the contestant's effort cost $c_c(q)$ increases with his performance, i.e. $c_c\left(q^L\right) < c_c\left(q^H\right)$.

Accordingly, the referee's expected payoff is represented by

$$E\left[\pi_r(q, z)\right] = (1 + E\left[H(v_r)\right]) S + G(q, z) B(z) - c_r - k_r(z), \qquad (6.1)$$

---

[11]   The referee's expected probability of re-assignment directly results from the probability distribution of the signal $v_r \in \left\{v_r^L, v_r^H\right\}$.

[12]   There is no obvious reason why a referee should incur a higher or lower effort cost from making honest versus deliberately wrong decisions.

while the contestant's expected payoff can be stated as

$$E\left[\pi_c\left(q,z\right)\right] = G\left(q,z\right)\left(V - B\left(z\right)\right) - c_c\left(q\right) - k_c\left(z\right). \tag{6.2}$$

Again, it will be assumed for simplicity that the contestant's participation constraints, namely

$$0 \leq \max\left\{E\left[\pi_c\left(q^L,m\right)\right], E\left[\pi_c\left(q^H,m\right)\right]\right\}$$

in a game with an honest referee and

$$0 \leq \max\left\{E\left[\pi_c\left(q^L,m^b\right)\right], E\left[\pi_c\left(q^H,m^b\right)\right]\right\}$$

in a game with a biased referee, will always be satisfied.

Let the game quality be indicated by $\phi\left(q,z\right)$, and suppose that it is an increasing function of both the contestant's and the referee's performance, so that $\phi\left(q^H,z\right) > \phi\left(q^L,z\right)$ and $\phi\left(q,m\right) > \phi\left(q,m^b\right)$.

The sports federation's goal is then to maximize its payoff while incentivizing the contestant to exert a high performance q^H and the referee to be honest (i.e. to elicit a high performance m). The sports association's payoff is given by

$$\pi_a = R\left(\phi\left(q^H,m\right)\right) - S - V, \tag{6.3}$$

where $R\left(\phi\left(q^H,m\right)\right)$ indicates the sports association's revenue from a contest as a function of the desired game quality $\phi\left(q^H,m\right)$. Thus, the sports association's optimization problem becomes:

$$\min_{S,V} S + V \quad s.t. \quad \phi\left(q,z\right) = \phi\left(q^H,m\right) \tag{6.4}$$

In other words, the sports association wants to minimize S and V while still attaining the high game quality $\phi\left(q^H,m\right)$.

In order to do so, the sports federation has to take both the contestant's and the referee's payoff functions into account, where, as in the previous model, all parties are assumed to be risk neutral. Note in this context that the contestant's winning probability $G(q, z)$ and the bribe $B(z)$ make the referee's and the contestant's payoff function interdependent, which in turn suggests an interdependence between the referee's and the contestant's incentives. It is important for the sports association to be aware of this interdependence because it means that the optimal incentive contract S to prevent sports corruption will not only depend on the referee's but also on the contestant's incentives. This is demonstrated in more detail below, where we will now study the optimal solutions for the association's optimization problem (6.4).

6.3   Optimal Incentivization Mechanism - Two Options

The sports association can attain the desired game quality $\phi(q^H, m)$ in two ways:

1. Option (1): The sports association incentivizes the contestant to exert a high effort and ensures that the contestant does not bribe the referee.

2. Option (2): The sports association incentivizes the contestant to exert a high effort and ensures that the referee rejects the contestant's bribe.

The difference between Option (1) and Option (2) lies in the target of incentivization (and hence of a federation's ultimate anti-corruption policies). Whereas Option (1) targets the contestant, Option (2) targets the referee.

Accordingly, the necessary conditions to be satisfied in the association's optimization problem vary across the two options, most likely leading to different optimal corruption prevention solutions.

Availing of the Lagrange Theorem allows us to solve the sports association's optimization problem for both options. The according technical derivations of the distinct solutions for Option (1) and Option (2) are presented in Appendix A.4 and Appendix A.5 respectively. Referring to these derivations, the following will only focus on the discussion of the solutions and what they imply for the optimal design of anti-corruption policies.

6.3.1   Option (1) - Targeting the Contestant

Optimal V   No matter which of the two options the sports association pursues, attaining the desired game quality $\phi\left(q^H, m\right)$ always requires the sports association to ensure that the contestant prefers expending a high effort over eliciting a low effort in a game with an honest referee. The particularity of Option (1), as a way of preventing sports corruption, is, however, that it must concurrently not be worthwhile for the contestant to bribe the referee.

An optimal avail of the winner's prize as an instrument of impulsion via Option (1) therefore requires the following two conditions to be satisfied:

(a) The contestant's expected payoff from exerting a low effort does not exceed his expected payoff from exerting a high effort in a non-manipulated game, i.e. $E\left[\pi_c\left(q^L, m\right)\right] \leq E\left[\pi_c\left(q^H, m\right)\right]$.

(b) The contestant's expected payoff from exerting a low effort and his expected payoff from exerting a high effort in a manipulated game does not exceed his expected payoff from exerting a high effort in a non-manipulated game, i.e. $\max\left\{E\left[\pi_c\left(q^L, m^b\right)\right], E\left[\pi_c\left(q^H, m^b\right)\right]\right\} \leq E\left[\pi_c\left(q^H, m\right)\right]$.

Condition (b) implies that there are two principal cases (and hence multiple possibilities) potentially providing an optimal solution for the winner's prize: Case 1, where $E\left[\pi_c\left(q^L, m^b\right)\right] > E\left[\pi_c\left(q^H, m^b\right)\right]$, and Case 2, where $E\left[\pi_c\left(q^L, m^b\right)\right] < E\left[\pi_c\left(q^H, m^b\right)\right]$.[13]

Yet, despite the initially apparent multiplicity of possible solutions, the comparison of the necessary conditions of all possibilities clarifies that there can indeed only be one optimal solution for the winner's prize V under Option (1). In particular, it appears to be optimal for the sports association to set the winner's prize at

$$V = \frac{c_c\left(q^H\right) - c_c\left(q^L\right)}{G\left(q^H, m\right) - G\left(q^L, m\right)}, \qquad (6.5)$$

where it just satisfies condition (a) above. In other words, the optimal winner's prize equals the contestant's cost of switching from a low to a high effort in a game with an honest referee relative to the associated gain in his probability of success.

At the same time, however, this solution requires that

$$\frac{c_c\left(q^H\right) - c_c\left(q^L\right)}{G\left(q^H, m\right) - G\left(q^L, m\right)} \leq \frac{G\left(q^L, m^b\right) B\left(m^b\right) + k_c\left(m^b\right) + c_c\left(q^L\right) - c_c\left(q^H\right)}{G\left(q^L, m^b\right) - G\left(q^H, m\right)}$$

$$(6.6)$$

holds for condition (b) to be satisfied as well. In other words, the optimal winner's prize must not exceed the contestant's cost of switching his strategy from simply exerting a legitimate high effort to bribing the referee and exerting a low effort relative to the associated gain in his probability of success.

Condition (6.6) in particular results from the fact that at the optimal winner's prize (6.5) the contestant's expected payoff from exerting a low effort exceeds his expected payoff from exerting a high effort in a game with

---

[13] For a detailed outline of the two cases, please refer to Appendix A.4.

a biased referee (i.e. Case 1 provides the optimal solution).[14] In this way, condition (6.6) determines the necessary condition for the contestant not to be interested in offering the referee a side contract (and hence for Option (1) to be pursuable).

Proposition 8 *The optimal winner's prize V in Option (1) is given by expression (6.5) and must satisfy condition (6.6).*

To understand more intuitively why at the winner's prize given by expression (6.5) the contestant attains a higher expected payoff from exerting a low effort than from exerting a high effort in a manipulated game, compare the contestant's incentives in the case where the referee is honest with his incentives where the referee is biased.

In a game with an honest referee, the contestant exerts a high effort if $E\left[\pi_c\left(q^L,m\right)\right] \leq E\left[\pi_c\left(q^H,m\right)\right]$, or equivalently, if

$$\frac{c_c\left(q^H\right) - c_c\left(q^L\right)}{V} \leq G\left(q^H,m\right) - G\left(q^L,m\right), \qquad (6.7)$$

is satisfied. In a game with a biased referee, the contestant would in turn only exert a high effort if $E\left[\pi_c\left(q^L,m^b\right)\right] \leq E\left[\pi_c\left(q^H,m^b\right)\right]$, or if

$$\frac{c_c\left(q^H\right) - c_c\left(q^L\right)}{V - B\left(m^b\right)} \leq G\left(q^H,m^b\right) - G\left(q^L,m^b\right), \qquad (6.8)$$

is satisfied.

Taking a closer look at conditions (6.7) and (6.8) clarifies that the contestant's incentives to exert a high effort vary with the referee's state of

---

[14] Put differently, because at the optimal winner's prize (6.5) $\max\left\{E\left[\pi_c\left(q^L,m^b\right)\right], E\left[\pi_c\left(q^H,m^b\right)\right]\right\} = E\left[\pi_c\left(q^L,m^b\right)\right]$, $E\left[\pi_c\left(q^L,m^b\right)\right] \leq E\left[\pi_c\left(q^H,m\right)\right]$ rather than $E\left[\pi_c\left(q^H,m^b\right)\right] \leq E\left[\pi_c\left(q^H,m\right)\right]$ must hold to prevent the contestant from bribing the referee. For the according technical proof, please refer to Appendix A.4.

bias. By assumption $G\left(q^H, m^b\right) - G\left(q^L, m^b\right) < G\left(q^H, m\right) - G\left(q^L, m\right)$, so that the right hand side (RHS) of condition (6.7) is larger than the RHS of condition (6.8), while for any bribe $B\left(m^b\right) \geq 0$ the left hand side (LHS) of condition (6.7) would never exceed the left hand side of condition (6.8).

Therefore, the contestant is more likely to exert a high effort, and thereby to contribute to a high game quality, in a game with an honest referee than in a game with a biased referee. This is due to the fact that a cheating contestant incurs the extra cost from bribery, lowering his incentives to concurrently exert a high effort.

This means that the sports association would have to provide the contestant with higher incentives to display a high effort in a manipulated game than in a non-manipulated game. While, in a game with an honest referee, it is sufficient for the sports association to set V such that

$$\frac{c_c\left(q^H\right) - c_c\left(q^L\right)}{G\left(q^H, m\right) - G\left(q^L, m\right)} \leq V \tag{6.9}$$

to incentivize high effort, the sports association would have to set V as high as

$$\frac{c_c\left(q^H\right) - c_c\left(q^L\right)}{G\left(q^H, m^b\right) - G\left(q^L, m^b\right)} + B\left(m^b\right) \leq V \tag{6.10}$$

in order to incentivize the contestant to exert a high effort in a game with a biased referee.[15]

If the sports association now sets V according to expression (6.5), so that it just satisfies condition (6.9), condition (6.10) will be left unsatisfied. Thus, at the optimal winner's prize V for Option (1) the contestant would never be willing to exert a high effort in a game with a biased ref-

---

[15] Note that for $B\left(m^b\right) \geq 0$ the LHS of condition (6.10) is unambiguously higher than the LHS of condition (6.9), while it seems that the higher the winner's prize V and the lower the bribe $B\left(m^b\right)$, the more likely it is that $E\left[\pi_c\left(q^L, m^b\right)\right] \leq E\left[\pi_c\left(q^H, m^b\right)\right]$ holds rather than $E\left[\pi_c\left(q^L, m^b\right)\right] \geq E\left[\pi_c\left(q^H, m^b\right)\right]$.

eree. From this follows that at the winner's prize given by expression (6.5)
$\max \left\{ E\left[\pi_c\left(q^L, m^b\right)\right], E\left[\pi_c\left(q^H, m^b\right)\right] \right\} = E\left[\pi_c\left(q^L, m^b\right)\right]$.

In any case, the optimal solution for Option (1) illustrates that, while
the winner's prize V must be high enough for the contestant to exert a high
effort in a game with an honest referee, it must not exceed a certain *upper*
limit, above which the contestant becomes interested in bribing the referee.[16]

Optimal S As Option (1) can only be pursued if the contestant refrains
from bribing the referee (i.e. if condition (6.6) holds), there is no need to
further incentivize the referee to reject a bribe in Option (1). The referee
would simply not be offered the opportunity of entering a side contract.

Therefore, in combination with the optimal winner's prize (6.5) it fully
suffices for the sports association to offer the referee an incentive contract
that merely satisfies the referee's participation constraint for supervising a
professional game honestly.[17] Thus, as the results in Appendix A.4 illustrate,
the optimal incentive contract offered to the referee in Option (1) is given
by:

$$S = \frac{c_r}{1 + p_r H\left(v_r^L\right) + (1 - p_r) H\left(v_r^H\right)} \tag{6.11}$$

Proposition 9 *The optimal incentive contract S offered to the referee in
Option (1) is given by expression (6.11).*

Summarizing, for Option (1) the sports association optimally sets the
winner's prize V according to expression (6.5). However, for Option (1)
to be pursuable, i.e. to attain the desired game quality $\phi\left(q^H, m\right)$ without
having to provide the referee with a corruption deterring incentive contract,
condition (6.6) must be satisfied to deter the contestant from bribing the

---

[16] Here, this upper limit is provided by the RHS of condition (6.6).

[17] Because at the optimal winner's prize (6.5) the contestant prefers to exert a high
rather than a low effort in a game with an honest referee, the referee's participation
constraint is technically given by $0 \le E\left[\pi_r\left(q^H, m\right)\right]$. Note, however, that an honest
referee is indifferent between Case 1 and Case 2, as $E\left[\pi_r\left(q^H, m\right)\right] = E\left[\pi_r\left(q^L, m\right)\right]$.

referee. If condition (6.6) indeed holds, it would suffice for the sports association to offer the referee an incentive contract S that satisfies the referee's participation constraint for honest behavior, i.e. to offer an incentive contract S according to expression (6.11). Otherwise, the sports association has to attend to Option (2), which will be discussed now.

6.3.2 Option (2) - Targeting the Referee

Optimal V As in Option (1), the sports association has to set the winner's prize at a level at which the contestant prefers expending a high effort over eliciting a low effort in a game with an honest referee. However, unlike in Option (1), this winner's prize must now be so high that the contestant has an incentive to bribe the referee. Hence, the optimal winner's prize under Option (2) must satisfy the following two conditions:

(a) The contestant's expected payoff from exerting a low effort does not exceed his expected payoff from exerting a high effort in a non-manipulated game, i.e. $E\left[\pi_c\left(q^L, m\right)\right] \leq E\left[\pi_c\left(q^H, m\right)\right]$.

(b) The contestant's expected payoff from exerting a high effort in a non-manipulated game falls either short of his expected payoff from exerting a low effort in a manipulated game and/or his expected payoff from exerting a high effort in a manipulated game, i.e. $E\left[\pi_c\left(q^H, m\right)\right] \leq \max\left\{E\left[\pi_c\left(q^L, m^b\right)\right], E\left[\pi_c\left(q^H, m^b\right)\right]\right\}$.

Again, the nature of condition (b) implies that there are two cardinal cases possibly providing the optimal solution for the winner's prize V that must be examined.

Nevertheless, studying each of the eligible distinct solutions subject to their necessary constraints clarifies that, similar to Option (1), there is in fact only one optimal solution for Option (2). While this optimal solution again requires to set the winner's prize V according to expression (6.5),

$$V = \frac{c_c\left(q^H\right) - c_c\left(q^L\right)}{G\left(q^H, m\right) - G\left(q^L, m\right)},$$

it now necessitates that

$$\frac{c_c\left(q^H\right) - c_c\left(q^L\right)}{G\left(q^H, m\right) - G\left(q^L, m\right)} \geq \frac{G\left(q^L, m^b\right) B\left(m^b\right) + k_c\left(m^b\right) + c_c\left(q^L\right) - c_c\left(q^H\right)}{G\left(q^L, m^b\right) - G\left(q^H, m\right)}$$
(6.12)

holds for condition (b) under Option (2) to be satisfied as well.[18] Thus, now the optimal winner's prize must not fall short of the contestant's cost of switching his strategy from simply exerting a legitimate high effort to bribing the referee and exerting a low effort relative to the associated gain in his probability of success.

Hence, the optimal winner's prize under Option (2) is the same as in Option (1), but now the contestant must be willing to bribe the referee. Condition (6.12) requires exactly that, given that at the optimal winner's prize (6.5) the contestant's expected payoff from exerting a low effort exceeds his expected payoff from exerting a high effort in a manipulated game.

Proposition 10 *The optimal winner's prize V for Option (2) is the same as for Option (1), i.e. given by expression (6.5), but it must now satisfy condition (6.12).*

At any rate, this optimal solution indicates that, conversely to Option (1), the winner's prize V must not fall short of a certain *lower* limit, for it to be worthwhile for the contestant to bribe the referee (and hence for Option (2) to be pursuable).

---

[18] Because the optimal winner's prize is again given by equation (6.5), it should not be surprising that condition (6.12) requires that $E\left[\pi_c\left(q^H, m\right)\right] \leq E\left[\pi_c\left(q^L, m^b\right)\right]$ holds rather than $E\left[\pi_c\left(q^H, m\right)\right] \leq E\left[\pi_c\left(q^H, m^b\right)\right]$.

Optimal S   Because in Option (2) the sports association indeed allows for the possibility - or rather by necessity ensures - that the referee is offered a side contract, the sports association now has to make sure that the referee dismisses such an enticement.

To find the most cost efficient way of doing so, it is important to recognize that the referee would be more inclined to enter a side contract if the contestant would exert a high effort in a manipulated game than if he would only exert a low effort. This is because the contestant's effort choice influences the probability of success of the deal, and thereby the probability with which the referee is going to be paid the bribe $B\left(m^b\right)$.[19] This implies that the sports association would have to offer a higher incentive contract S to the referee, if the contestant would exert a high effort in a game with a biased referee.

To see this, observe that if $\max\left\{E\left[\pi_c\left(q^L,m^b\right)\right],E\left[\pi_c\left(q^H,m^b\right)\right]\right\}=E\left[\pi_c\left(q^H,m^b\right)\right]$, the sports association would have to offer an incentive contract

$$S=\frac{G\left(q^H,m^b\right)B\left(m^b\right)-k_r\left(m^b\right)}{\left(1-p_r\right)\left(H\left(v_r^H\right)-H\left(v_r^L\right)\right)},\tag{6.13}$$

i.e. that just satisfies $E\left[\pi_r\left(q^H,m^b\right)\right]\leq E\left[\pi_r\left(q^H,m\right)\right]$, to prevent the referee from entering a side contract with the contestant. If it is, however, the case that $\max\left\{E\left[\pi_c\left(q^L,m^b\right)\right],E\left[\pi_c\left(q^H,m^b\right)\right]\right\}=E\left[\pi_c\left(q^L,m^b\right)\right]$ a sports association would only have to offer an incentive contract

$$S=\frac{G\left(q^L,m^b\right)B\left(m^b\right)-k_r\left(m^b\right)}{\left(1-p_r\right)\left(H\left(v_r^H\right)-H\left(v_r^L\right)\right)},\tag{6.14}$$

that just satisfies $E\left[\pi_r\left(q^L,m^b\right)\right]\leq E\left[\pi_r\left(q^H,m\right)\right]$.

Because the optimal winner's prize V determined by expression (6.5)

---

[19]   In mathematical terms, because $G\left(q^H,m^b\right)>G\left(q^L,m^b\right)$, $E\left[\pi_r\left(q^H,m^b\right)\right]>E\left[\pi_r\left(q^L,m^b\right)\right]$.

leaves condition (6.10) unsatisfied so that the referee would only be able to attain the expected payoff $E\left[\pi_r\left(q^L, m^b\right)\right]$, expression (6.14) represents the optimal incentive contract in Option (2). Clearly, because $G\left(q^H, m^b\right) > G\left(q^L, m^b\right)$, the optimal incentive contract S described by expression (6.14) is lower than that described by expression (6.13).

Proposition 11 *The optimal incentive contract S offered to the referee in Option (2) is given by expression (6.14).*

The solutions (6.13) and (6.14) highlight the fundamental line of argument, namely that the optimal incentive contract S for the referee depends on the contestant's optimal effort choice, which in turn depends on the prize V the contestant is awarded with if he wins the contest.

As a result, in the attempt to provide the referee with efficient incentives, it would be counter-productive to raise V to a level where it would also satisfy condition (6.10), because this would redundantly raise the referee's incentives for corruption and hence the cost of providing the referee with honest incentives.

Proposition 12 *Raising V to a level satisfying condition (6.10) would redundantly raise the cost of incentivizing the referee to be honest.*

Summarizing, for Option (2) the sports association sets the optimal winner's prize at the same level as in Option (1) (i.e. according to expression (6.5)). However, for Option (2) to be pursuable, this winner's prize must satisfy condition (6.12). If condition (6.12) is indeed satisfied, the sports association has to annihilate the referee's enticement of entering a side contract offered by the contestant. To do so, the sports association optimally offers the referee an incentive contract S according to expression (6.14).

6.3.3 Option (1) versus Option (2)

The optimal solutions discussed above clarify that, independent of the option pursued, the optimal prize level V is always set equal to the contestant's cost of switching his strategy from exerting a low to a high effort relative to the resulting gain in the probability of success (i.e. according to expression (6.5)). Merely the optimal incentive contract S to be offered to the referee differs across the two options.

Yet, assuming that the referee is corruptible at all, i.e. assuming that $E\left[\pi_r\left(q^L, m^b\right)\right] > 0$, it is also clear that offering an incentive contract that just satisfies the referee's participation constraint is always the more cost efficient option. In other words, Option (1) is always the preferable solution for attaining the desired game quality $\phi\left(q^H, m\right)$.

Proposition 13 *Option (1) is the preferable solution for attaining the desired game quality $\phi\left(q^H, m\right)$.*

However, this requires that the pursuit of Option (1) subject to its necessary condition is indeed feasible. If at the optimal winner's prize (6.5) the necessary condition of Option (1) (i.e. condition (6.6)), is not satisfied, the sports association has to attend to Option (2). This happens if due to the nature of the game, the contestant's cost of exerting a high instead of a low effort relative to its gain in the winning probability is exceedingly high, or conversely, if the contestant's expected cost of switching from merely exerting a legitimate high effort to bribing the referee and exerting a low effort relative to its gain in the winning probability is exceedingly low.

Re-arranging equation (6.5) and plugging it into condition (6.6), simplifies condition (6.6) to:

$$V \leq \frac{G\left(q^L, m^b\right) B\left(m^b\right) + k_c\left(m^b\right)}{G\left(q^L, m^b\right) - G\left(q^L, m\right)} \tag{6.15}$$

Thus, as long as the winner's prize necessary to incentivize the contestant to exert a high rather than a low effort in a game with an honest referee does not exceed the contestant's expected cost from corruption relative to its associated gain in the winning probability, Option (1) can (and should) be pursued.

Proposition 14 *If the winner's prize level V set according to expression (6.5) satisfies condition (6.15) (or alternatively condition (6.6)), the sports association should pursue Option (1). Otherwise it should pursue Option (2).*

This analysis therefore provides two important insights. First, the sports association imperatively has to account for the interdependencies between the contestant's and the referee's optimal behavior in order to implement cost efficient (and hence profit maximizing) incentivization mechanisms while yielding a high game quality $\phi\left(q^H, m\right)$. Second, sports associations should always start by targeting the contestant rather than the referee, or, put differently, by making Option (1) pursuable rather than Option (2).[20]

For this purpose, it is important that sports associations are well informed about the explicit determinants of the contestant's and the referee's incentives for collusion.

6.4   A Contestant's Incentives to Bribe the Referee

The contestant's incentives to bribe a referee are essentially described by condition (6.15), which is replicated here for completeness.

$$V \leq \frac{G\left(q^L, m^b\right) B\left(m^b\right) + k_c\left(m^b\right)}{G\left(q^L, m^b\right) - G\left(q^L, m\right)}$$

---

[20]   This result in a sense controverses Kingston's (2007) conclusion that it would be optimal to punish only the recipient of a bribe in order to deter "*parochial*" corruption.

The more likely condition (6.15) is satisfied, the lower the contestant's incentives to bribe the referee and hence the more cost efficient it is for a sports association to attain a high game quality $\phi\left(q^H, m\right)$. Remembering that

$$V = \frac{c_c\left(q^H\right) - c_c\left(q^L\right)}{G\left(q^H, m\right) - G\left(q^L, m\right)}$$

suggests the following determinants of the contestant's bribery incentives.

Clearly, the higher the contestant's cost of switching from a low to a high effort $c_c\left(q^H\right) - c_c\left(q^L\right)$, the higher the winner's prize necessary to incentivize the contestant to exert a high effort in a non-manipulated game. This means that his incentive to offer the referee a side contract linearly increases with the difference in effort costs. The reason is that the cost of legitimately as opposed to illicitly trying to win the contest via bribery would be relatively high.

Additionally, the contestant's incentives to bribe the referee increase in the gain in the contestant's probability of success attainable from bribing the referee, i.e. in $G\left(q^L, m^b\right) - G\left(q^L, m\right)$. This is because the effectiveness and hence the expected gain from bribery as opposed to legitimately trying to win the contest is simply higher.

Conversely, the higher the gain in the winning probability $G\left(q^H, m\right) - G\left(q^L, m\right)$ resulting from exerting a high effort rather than a low effort in a non-manipulated game, the lower the contestant's incentives for sports corruption because the expected gain from exerting a high effort legitimately would be higher. As a result, the winner's prize necessary to get the contestant to exert a high effort in a game with an honest referee is also lower.

Furthermore, it is not surprising that the contestant's incentives to bribe the referee are inversely correlated with the miscellaneous costs $k_c\left(m^b\right)$ incurred from doing so.

Equivalently, the higher the expected bribe $G\left(q^L, m^b\right) B\left(m^b\right)$ to be paid to the referee, the lower the contestant's incentives to bribe the referee. This is because a higher expected bribe increases the contestant's expected cost from corruption relative to its expected gains.

Proposition 15 *The contestant's incentives to bribe the referee increase in*

(a) the winner's prize V,

(b) the switching costs $c_c\left(q^H\right) - c_c\left(q^L\right)$ from exerting a high rather than a low effort,

(c) and the gain in the winning probability $G\left(q^L, m^b\right) - G\left(q^L, m\right)$ from bribing the referee.

Conversely, the contestant's incentives to bribe the referee decrease in

(a) the gain in the winning probability $G\left(q^H, m\right) - G\left(q^L, m\right)$ resulting from exerting a high effort rather than a low effort in a game with an honest referee,

(b) the miscellaneous costs $k_c\left(m^b\right)$ from corruption,

(c) and the expected bribe $G\left(q^L, m^b\right) B\left(m^b\right)$ to be paid to the referee.

6.5   A Referee's Incentives to Accept a Bribe

Now, to assess the determinants of the referee's incentives for corruption in more detail, infer from expression (6.14), that for an optimal winner's prize V set according to expression (6.5), the referee would reject any bribe offered by the contestant as long as

$$S \geq \frac{G\left(q^L, m^b\right) B\left(m^b\right) - k_r\left(m^b\right)}{\left(1 - p_r\right)\left(H\left(v_r^H\right) - H\left(v_r^L\right)\right)}. \tag{6.16}$$

The more likely condition (6.16) is satisfied, the lower the referee's incentives to accept the contestant's bribe. This suggests the following determinants of the referee's incentives for corruption.

Not surprisingly, the referee's incentives to enter a side contract increase with the expected bribe $G\left(q^L, m^b\right) B\left(m^b\right)$ he receives in return for being biased in favour of the contracting contestant, as it directly increases the referee's expected payoff from corruption.

Besides, the referee's corrupt incentives increase in the probability p_r that the sports association receives a signal $v_r = v_r^L$ although the referee elicited a high performance m. This is because, if p_r is high, it is less likely that his honesty is rewarded in terms of a higher probability of being assigned to the next game.

On the other hand, the referee's incentives to agree to a side contract decrease with the incentive contract S. The higher S, the higher the expected loss from corruption in the future due to the lower likelihood of being re-assigned to another game resulting therefrom.

At the same time, similar to the contestant's bribery incentives, the referee's incentives for corruption decrease with the miscellaneous costs $k_r\left(m^b\right)$ incurred from corruption.

Furthermore, the higher the gain in the probability of re-assignment in the future from a high signal v_r^H being reported rather than a low signal v_r^L, i.e. the higher $H\left(v_r^H\right) - H\left(v_r^L\right)$, the lower the referee's incentives for corruption. The reason is that the benefit from being honest and hence the opportunity cost from corruption would be higher.

Proposition 16 *The referee's incentives to accept a bribe increase in*

(a) the expected bribe $G\left(q^L, m^b\right) B\left(m^b\right)$,

(b) and the probability p_r that the sports association observes a signal v_r^L despite a high refereeing performance.

Conversely, the referee's incentives to accept a bribe decrease in

(a) the fee S paid by the sports association per game supervised,

(b) the miscellaneous costs $k_r\left(m^b\right)$ from corruption,

(c) and the gain in the probability of being re-assigned to another game in the future $H\left(v_r^H\right) - H\left(v_r^L\right)$ from having a high signal v_r^H rather than a low signal v_r^L being reported.

Chapter 7

Anti-Corruption Policy Suggestions

With the insights gained from the preceding analysis let us now derive and assess a range of possible anti-corruption policies. As the bribery of a referee requires the agreement of both the contestant and the referee, the deterrence of such delinquent activities necessitates their associated costs to be somehow elevated above their benefits for at least one of the two parties.[1]

Yet, the economic conception is that society should not try to deter culpable activities at all costs. Instead, a rational implementation of anti-corruption policies requires that they are only availed of as long as their social marginal costs do not exceed their social marginal benefits. This approach leads to the widely supported opinion in the economic literature that the socially optimal level of delinquent acts may not be zero, implying that a

---

[1] For more information on the economics of crime deterrence, see Becker's (1968) seminal work on crime and punishment.

certain level of sports corruption may be acceptable.[2]

However, the particularities of the sports industry might suggest otherwise. Given that the first discovered corruption case in the sports industry can already cause a tremendous marginal social damage, as it not only hurts the image of those directly involved but also that of the entire sports society, it could indeed be argued that the level of corruption in the sports industry is optimally reduced to zero.[3]

The following presents some suggestions for anti-corruption measures that might help to do so, while assessing their particular usefulness with regard to the pursuit of Option (1) and Option (2). After considering a few more general anti-corruption policies, more specific contest design and regulatory policies are discussed. Finally, the Bavarian Draft Bill will be reconsidered to study its economic effect.

7.1   General and Institutional Policies

7.1.1   Responsible Media Coverage

A first step into the right direction towards an efficient corruption prevention culture in professional German sports leagues can already be made by the public press. A responsible media coverage of ongoing investigations on sports corruption cases, promoting the perception that sports corruption will be discovered and punished accordingly, can have an efficient corruption deterring effect.[4]

The transparent provision of information on sports corruption cases enhances the scrutinization of suspicious behaviors by the public. Such public pressure in turn leads to a higher perceived risk of corruption for contestants and referees, which increases both cost parameters k_c and k_r. In this way,

---

2   Ehrlich (1996, p. 51), Becker (1968, pp. 180 et seq.)
3   Koch and Maennig (2007, p. 53)
4   Andersen (2006, p. 92), Koch and Maennig (2007, p. 53)

a responsible media coverage can not only lower the referee's but also the contestant's incentives for corruption, which would simplify the pursuit of Option (1).

However, a responsible media coverage also entails only approved information on possible sports corruption cases to be published. This is especially important because unwarily caused hypes by the spread of merely speculative information might lead to a seminal "*corruption culture*". The higher the level of sports corruption perceived by those involved in the sports industry, the lower will be their moral threshold to engage in it,[5] suggesting that corruption in essence becomes a self-fulfilling prophecy.[6]

Even more unacceptable is, however, the seemingly strong interest of sports associations to keep the problem of sports corruption under wraps. The fact that four speakers, who initially firmly accepted an invitation to discuss the state of affairs in the International Soccer Federation (FIFA) at the "*Play the Game*" conference in 2002, for dubious reasons suddenly had to cancel their attendance shortly before the conference, as well as the allegation of the International Volleyball Federation (FIVB) in its response to such an invitation in 2005 that discussing the issue of sports corruption in the federation would be "*illegal*" exemplify the highly equivocal information politics of some sports federations.[7]

One of the reasons for why it is so difficult for sports associations to self-reflect and open itself up for criticism is that the sports society cultivates a strong sense of cohesion towards the public society. Not to criticize or damage its members in any way seems to be an unwritten law in the sports society. This exclusive society not only includes officials, referees and

---

5   Bardhan (2006, pp. 344 et seq.), Andvig and Moene (1990, pp. 71 et seq.)
6   Warren and Laufer (2009) explain the same phenomenon with respect to labeling countries with corruption indices.
7   Andersen (2006, p. 80)

contestants but among others also sponsors and the media.[8]

A transparent media coverage would constitute a betrayal of some members because it might lower the returns of investments for sponsors as a result of a potentially impaired image of the sports association. In fact, media conglomerates might themselves face a conflict of interest in its media coverage, as some of them invest a large amount of resources to attain the broadcasting rights of sports events, for instance. Therefore, by providing the public with transparent information on sports corruption, media conglomerates risk a severe incursion in the return of their own investments as well.[9]

Hence, despite the potential effectiveness of a responsible media coverage as a corruption prevention policy from an economic perspective, it is unlikely to be patronized by the sports society.

7.1.2 Independent Investigating Institution

A possibly less objected policy might be the establishment of an independent (international) investigating institution that exclusively focuses on corrupt endeavors in the sports society.[10] While an international institution fighting doping delinquencies in sports, namely the World Anti-Doping Agency (WADA), has already been established, a complementary companion institution tackling the problem of sports corruption is still missing.

The establishment of such an institution (if not at an international level, at least at the national level) would raise the probability of detection for sports defrauders and constitute a strong signal by all German sports associations for their commitment towards controverting sports corruption in Germany.

As a result, the expected cost incurred from corruption for both referees

---

[8]   Andersen (2006, pp. 83 et seq.)
[9]   Mason et al. (2006, pp. 67 et seq.)
[10] Maennig (2005, pp. 211 et seq.), Koch and Maennig (2007, p. 53)

and contestants, i.e. both k_r and k_c, would increase, so that an independent investigating institution would also facilitate the pursuit of Option (1).

7.1.3 Code of Conduct

A relatively simple measure to help deterring sports corruption is also the integration of a clear code of conduct with a special focus on sports corruption in the bylaws of all sports associations.[11]

Such a code of practice, specifically outlining and defining undesirable behavior, would be beneficial in two aspects. On the one side, it would allow to specifically measure the degree of an individual's misconduct, which is necessary to design an efficient scale of penalties. On the other side, it would also serve as a strong signal for a sports association's contempt for corrupt endeavors and its dedication to fight them.[12]

Therefore, a clear behavioral code would most likely not only raise an individual's moral cost incurred from corruption, but also elevate the expected penalty as perceived by potential sports defrauders. In this way, a code of conduct would again lead to an increase in k_r and k_c, causing an alleviation in the corruption incentives of both, the referee and the contestant. Hence, this is also a possible anti-corruption policy that would help to pursue Option (1).

Nevertheless, unlike for doping delinquencies,[13] such a behavioral code focusing on sports corruption, as it is defined here, is still missing in the bylaws of most sports associations. This again underlines the common negligence of sports corruption in Germany. In fact, considering how effortless it would be to manifest such an anti-corruption code of conduct in the bylaws, once more indicates the apparent intent of most sports associations to play

---

[11]   Andersen (2006, p. 91), Maennig (2005, p. 216)
[12]   Koch and Maennig (2007, p. 53)
[13]   cf. DFB (2010c) and DHB (2009a)

down the severity of the problem of sports corruption in Germany.[14]

7.1.4   System of Checks and Balances

The importance of the structure of sports associations with respect to its corruption deterring effect must not be underestimated either. To minimize the level of sports corruption, a sports association has to provide for efficient incentives at all levels in its hierarchy.[15] In this sense, decentralizing the hierarchy structure of associations and separating the powers of control into a system of checks and balances can be of much help to reduce the level of sports corruption therein.[16]

Especially when it comes to the assignment of referees to professional games, their performance evaluation afterwards or the investigation of potential corruption cases, it must be ensured that the upper layers in the hierarchy (i.e. the controllers) are not corrupt themselves. Without doubt, the higher the percentage of corrupt controllers, the easier it is for those being monitored to collude with their supervisors, which in turn lowers their probability of detection.[17]

A referee might for instance agree to share the bribe $B\left(m^b\right)$ paid by the contestant with those evaluating his performance if, despite the necessarily poorer performance resulting from his bias, they report a high signal v_r^H rather than a low signal v_r^L on the referee's performance. This way, the referee would be able to retain a high expected probability of re-assignment despite his corrupt endeavors.

If in the form of a system of checks and balances multiple veto powers in

---

[14]   Peer (2009)

[15]   For more information on the problem of motivation in organisations, see Jost (2000a, pp. 177 et seq.).

[16]   Bardhan (2006, p. 346), Mason et al. (2006, p. 66); see also Shleifer and Vishny (1993) arguing that the structure of an institution is an important determinant of corruption.

[17]   Andvig and Moene (1990, p. 75)

the decision making process regarding corruption prone issues are instituted, however, controllers are not as susceptible to such bribery attempts any more. In any case, the referee would have to bribe multiple individuals, which would require a larger share of his bribe money received from the contestant.

It follows that, instituting checks and balances in the hierarchy of sports associations can have an efficient deterring effect for sports defrauders. Because this anti-corruption policy also increases the probability of detection, it not only targets the referee's but also the contestant's incentives. Thus, it conveniently helps to prevent corruption via Option (1) as well.

7.1.5 Whistleblowing System

In light of the general difficulty of detecting sports corruption, as involved parties typically devote a considerable amount of resources towards its secrecy, the implementation of a whistleblowing system can be another effective measure of raising the probability of detection and hence the expected cost from corruption. Some other industries have already successfully implemented such a system that ensures protection and anonymity to those (i.e. the "*whistle blowers*"), who report inside information on illicit behaviors by others.[18]

To further enhance the incentives of insiders, such as colleague referees or other contestants, to report such delinquent behavior, sports associations could even think about providing extra rewards to whistle blowers, if such inside information later leads to the successful discovery of a corruption case.

However, there is always the danger that such a system is misused by individuals who would like to calumniate colleagues or other contestants, maybe even to withdraw the attention of investigating authorities from them-

---

[18] Andersen (2006, p. 91), Argadona (2003, p. 261), Bardhan (2006, p. 346)

selves.[19]

7.1.6 Anonymous Help Organization

Sometimes, referees are forced to agree to a side contract as a result of difficulties arising in their personal lives. For instance, referees facing financial straits are particularly vulnerable to corruption. Thus, referees, who are addicted to gambling, for example, can become easy targets for potential bribers. Besides, it is also not uncommon that bribers start to threaten unwilling referees with physical violence so that referees in the end might only agree to manipulating games to preserve the safety of their own and their families' lives.[20]

Sports associations would therefore be well advised to create an anonymous help organization that referees with such personal problems can turn to.[21] This would relieve some of the possibly externally imposed pressures on referees, which in turn reduces their susceptibility to corruption. Yet, such an institution would only facilitate deterrence of corruption via Option (2).

7.2 Contest Design and Regulatory Policies

7.2.1 Technological Assistance

One mechanism, that has already been instituted in some sports and is more and more frequently discussed in others, is the introduction of video evidence and other technological aids as an assistance for referees to verify uncertain calls and/or to raise the transparency of their decisions.[22]

On the one hand, because the referee can verify the validity of his call in

---

[19] Maennig (2005, pp. 211 et seq.)
[20] Biermann and Wulzinger (2008)
[21] Voigt (2009)
[22] Video evidence is, for instance, already being used in ice hockey or American football.

case of doubt, the introduction of video evidence leads to fewer unconsciously mistaken calls. Now, if fewer unconscious errors are made, the actual performance of contestants is evaluated more accurately. This in turn implies that video evidence as an assistance would increase the gain in the probability of success from exerting a high effort as opposed to a low effort. In other words, measures of technological assistance would increase $G\left(q^{H}, m\right) - G\left(q^{L}, m\right)$ by raising $G\left(q^{H}, m\right)$ and reducing $G\left(q^{L}, m\right)$. It follows that the optimal winner's prize V necessary to incentivize the contestant to exert a high effort in a game with an honest referee could be lowered.

On the other hand, by increasing the transparency of decisions, the introduction of video evidence could reduce the referee's discretionary power, thereby making it harder for him to enforce deliberately wrong calls.[23] This would in turn reduce a contestant's probability of success from bribing the referee, meaning that under the optimal winner's prize of Option (1) and (2), the contestant would expect the success of the deal to be less likely. Thus, assisting technological measures lower the expected bribe $G\left(q^{L}, m^{b}\right) B\left(m^{b}\right)$ payable to the referee and therefore also the contestant's expected cost from corruption.

It follows that the introduction of video evidence does not necessarily help to lower the contestant's incentives to bribe the referee. Nevertheless, it will reduce the referee's incentive to accept a bribe because of the lower likelihood of success. In this way, technological measures still constitute a viable anti-corruption policy for Option (2).

Yet, despite the evident corruption deterring potential of such technological aids and the required technological possibilities being available, some

---

[23] Part of the evidently high level of sports corruption in the International Handball Federation (IHF) is often ascribed to the fact that referees have too much discretionary power. Especially the referee's entirely subjective assessment of passive play is often criticized. (Hellmuth I. and Ewers C. (2009)) For this reason, the former basketball referee Willy Bestgen suggests to introduce a shot clock in handball, similar to the 24-second shot clock in basketball. (Tuneke (2009))

sports associations, such as the DFB, vehemently oppose its institution. With the help of micro sensors built into soccer balls and shin pads communicating with a three dimensional tracking system plus the use of video evidence, arising uncertainties regarding the confirmation of goals, offside calls or any possibly outcome decisive calls for that matter, could immensely be reduced.[24]

The leading arguments for the DFB's resistance are, on the one hand, that the installment of technological aids would undermine a referee's authority in his instant factual decision making ability and, on the other hand, that it would disrupt the flow of the game.[25] However, both arguments seem to be irrational compared to the benefits of such technological assistance.

Human beings inevitably make mistakes, and neither the contestants nor the public would expect otherwise. Allowing a referee to doubt his call and possibly to correct a mistake should only strengthen his competence in the eyes of the contestants and the public because, at least in the end, he made a correct call. Furthermore, one could easily limit the technology's disruption of the game by providing the contestants with a limited number of objection rights during the game, leading to a review of a referee's call via the technological assistance. If after review it turns out that the referee's decision was right, the objecting contestant loses one of them. Otherwise, the contestant gets to keep the objection right.[26]

7.2.2 Skill Enhancement of Referees

A similar effect as that of the introduction of technological aids would be attained by enhancing the skills of professional referees. Because the en-

---

[24]   Koch and Maennig (2007, p. 55)
[25]   Koch and Maennig (2007, p. 55)
[26]   Koch and Maennig's (2007, footnote 23) suggestion that the right to initiate a review of decisions via technological assistance should only be reserved to the referee is impracticle for corruption prevention purposes because a biased referee would never himself initiate a review of a deliberately wrong call.

hancement of the referees' skills would again lead to a more accurate evaluation of the actual performance of the contestants, a contestant's gain in the probability of success from switching from a low to a high effort will increase.

There are several possible methods for improving the skills of professional referees. Sports associations could, for instance, raise the number of required annual skill enhancing training sessions and/or the quality of such sessions. Another way would be to raise the overall quality standard a referee must fulfill to be allowed to supervise professional games.[27] Such an elevation of the quality standard would have the positive side effect of increasing a referee's cost of corruption k_r, because a referee would have more to lose in case of conviction. Yet, this is of course only an option up to the extent that there are enough referees that fulfill this elevated standard to satisfy the need for referees.

In any case, it is important that the skill enhancement process is consistent across all listed referees. This is because an asymmetric skill enhancement among referees might in fact lead to an increased level of sports corruption. If the skill level across referees differs significantly, a more skilled referee would be able to built up a superior reputation relative to a less skilled referee. This in turn means that the more skilled referee is likely to have a higher probability of being re-assigned to another game in the future, even after a low signal is reported by the performance evaluators on a so far highly reputed yet presently corrupt referee. An eminent skill gap across referees might therefore actually increase the incentives of a referee, who already managed to build up a superior reputation relative to other listed referees, to enter a side contract.[28]

Unlike the introduction of technological assistance, however, the skill

---

[27] This presumes that there are distinct skill elements that can be measured and ranked.

[28] Koch and Maennig's (2007, p. 54) suggestion of lowering the retirement age of referees in professional leagues might therefore also be worth thinking about.

enhancement of referees does not hinder them in any way to continue to enforce deliberately mistaken calls in the same way as before. Accordingly, the probability of success of exerting a low effort in a game with a biased referee $G\left(q^L, m^b\right)$, and hence the expected bribe receivable by the referee, will (if at all) not be lowered by as much. Thus, the contestant can not necessarily expect a reduction in the bribe payable, so that the skill enhancement of referees would, opposite to technological aids, also assist the pursuit of Option (1).

7.2.3   Efficient Performance Evaluation

Another possible anti-corruption measure that should be considered by sports associations is to increase the accuracy of the performance evaluation of referees. To do so, sports associations have to improve the performance evaluation methods so that the probability p_r of a wrong signal being reported decreases. This would lead to a higher expected probability $E\left[H\left(v_r\right)\right] = p_r H\left(v_r^L\right) + (1 - p_r) H\left(v_r^H\right)$ of being re-assigned for an honest referee. In this way, a referee's opportunity cost of corruption would be increased.[29]

However, such anti-corruption policies refining the expected re-assignment probability of referees only targets the referee and not the contestant. Thus, they are not suitable for enabling the pursuit of Option (1). Furthermore, such policies would only be effective if the controllers (the performance evaluators) are not corrupt, which further stresses the importance of the institution of checks and balances in the hierarchy structure of sports associations, as proposed earlier.

7.2.4   Job Rotation

The corruption convicted soccer referee Robert Hoyzer once mentioned that one of the pleasant aspects of being a sports referee is the establishment of

---

[29]   Remember here that $\Pr\left(v_r^L \mid m\right) = p_r$, $\Pr\left(v_r^H \mid m\right) = 1 - p_r$, $\Pr\left(v_r^L \mid m^b\right) = 1$, and $\Pr\left(v_r^H \mid m^b\right) = 0$.

many new acquaintances and friendships.[30] However, such close relationships inherently foster sports corruption, as they provide for the necessary amount of trust required for collusions to occur. Thus, another mechanism that might be useful to obviate sports corruption is to increase the job rotation among performance evaluators and among referees.

Among Performance Evaluators If the tenure of performance evaluators is shortened, it is less likely that such long-term relationships between referees and performance evaluators can be established. Staff rotation makes the behavior of evaluators less predictable for referees, which elevates the risk and thereby lowers the likelihood of collusion between performance evaluators and referees.[31] This would additionally enhance the accuracy of the performance evaluation as mentioned above.

Anyhow, increasing the job rotation among performance evaluators only affects the referee's incentives to enter a side contract and is thereby only useful for pursuing Option (2).

Among Referees An increased staff rotation among referees would in turn reduce a contestant's incentives to collude with a referee because the elevated uncertainty about the referee's behavior increases the contestant's risk of detection. Thus, not only should a sports association avoid games of a particular contestant being repeatedly supervised by the same referee, but it should also continuously interchange the pairings of referees in those types of sports, where groups of referees are assigned to professional matches. The German Basketball Federation (DBB) already does so effectively.[32] However, professional handball matches, for instance, are still commonly supervised by fixed pairs of referees.[33]

---

[30] Spiegel (2005)
[31] Abbink (2004, p. 888), Bardhan (1997, p. 1338)
[32] Voigt (2009)
[33] Hellmuth and Ewers (2009)

Because a more frequent job rotation among referees lowers a contestant's incentives to offer a referee a side contract, it would facilitate pursuing Option (1). However, to allow for such job rotation, it might be necessary to increase the number of referees eligible to supervise games at the professional level.

7.2.5 Number of Referees

Number of Eligible Referees Yet, increasing the number of eligible referees can also become an anti-corruption policy itself, as it would increase the competition among referees to be re-assigned to another game in the future.

If the competition between referees to be re-assigned to the next game is high, entering a side contract (and hence having a low signal v_r^L reported) would lower the probability of re-assignment by more than if this competition was weak. Thus, increasing the number of referees eligible to supervise professional games might actually increase the referee's expected cost from corruption.

However, a sports association should not exceedingly increase the number of listed referees because, if so, the opposite could in fact occur. If the number of listed referees reaches a level, where the competition among referees becomes fierce, the probability of re-assignment for a referee is already so low that the future cost of entering a side contract would only be minor as well. Referees would simply not have much to lose by accepting a bribe. In any case, increasing the number of eligible referees would also only be helpful for pursuing Option (2).

Number of Referees per Game Increasing the number of referees required to supervise a game might instead be a more effective corruption deterring measure.[34] The more referees are responsible for the supervision

---

[34] Koch and Maennig (2007, pp. 54 et seq.), Maennig (2005, p. 214)

of a particular game, the harder it is for one of these referees to enforce corrupt calls and thereby to influence the outcome of a game.

Because raising the required number of referees per game in this way would have the effect of lowering the success of an illicit deal (thereby reducing the likelihood that the contestant has to pay the bribe) one might be inclined to argue that this policy is also only helpful for preventing corruption via Option (2). However, one could also argue differently; in order to attain the same gain in the probability of success from bribery the contestant would have to bribe multiple referees. This increases the cost of closing the deal for the contestant, so that increasing the number of referees supervising a game could also be useful for Option (1).

7.2.6 Secrecy of Referee Assignment

One reason why it is often fairly easy for contestants to get in contact with the assigned referees is that in some sports referee assignments are announced long before the actual game. This makes it particularly easy for potential bribers (i.e. contestants) to know who to attend to in order to manipulate a game. Moreover, it provides bribers with a lot of time to close the deal. For this reason, the professional German Basketball League (DBL, hereinafter also referred to as "Basketball Bundesliga") has now started to keep the referee assignments secret until shortly before the game.[35]

However, it is still common in some sports that the hosting contestant organizes the accommodations and ensures the well-being of the assigned referee.[36] In the Handball Bundesliga, for example, it is not unusual for referees to be picked up from the airport, to be accommodated in nice hotels, to be invited to dinner, and occasionally to be taken on a city tour.[37] It is not surprising that under these circumstances assigned referees are occasionally

---

[35] Hellmuth and Ewers (2009), Voigt (2009)
[36] Focus (2009)
[37] Hellmuth and Ewers (2009)

offered additional enticements in various forms such as alcohol, food, women or money.[38]

By making referees organize their travel arrangements and accommodations themselves,[39] while keeping referee assignments secret until shortly before the game, one could significantly lower the chances of collusion between contestants and referees.[40] This would lead to an increase in searching costs and the costs of closing a deal, causing the contestant's cost of corruption k_c to increase.

As a result, this policy clearly facilitates the pursuit of Option (1). Of course, the larger the pool of eligible referees that could be assigned to a professional game, the more effective this policy will be.

7.2.7 Referee's Remuneration

Another anti-corruption policy that has been frequently discussed in the corruption literature in various contexts is to increase the remuneration of potential bribe recipients.[41] The argument is that raising their remuneration above their reservation wage (i.e. the wage they could earn elsewhere for comparable activities) increases their opportunity cost of corruption because the value of remaining in a job is elevated.[42]

However, as Sosa (2004) argues, this desired effect depends on the risk preferences of the recipient (i.e. the referee) and his salary after conviction, which might be difficult to identify.[43] Moreover, the inverse correlation between corruption and wages is only limited in its empirical support. While

---

[38]   Stern (2009)
[39]   Again, this is according to Voigt (2009) already common practice in the Basketball Bundesliga.
[40]   Schulze (2009), Koch and Maennig (2007, p. 55)
[41]   cf. Besley and McLaren (1993), Chand and Moene (1997)
[42]   Besley and McLaren (1993, pp. 120 et seq.), Maennig (2005, p. 216)
[43]   While risk neutrality was assumed in the model presented in Chapter 6, Sosa (2004, pp. 600 et seq.) assumes a decreasing risk aversion as income increases, so that he actually finds that corruption may even increase if wages are increased.

Goel and Rich (1989) find that in the United States higher salaries do indeed lead to a lower level of corruption in the public sector, Van Rijckeghem and Weder (2001) find contradictory econometric results.

Furthermore, although the DHB, for instance, only pays its referees a less remarkable amount of €500.- per game supervised in the first Handball Bundesliga and €300.- per game supervised in the second Handball Bundesliga,[44] the DFB already pays its referees a fairly high fixed fee per game supervised, as is shown in the table below.

	1. Bundesliga	2. Bundesliga
Referee	€3068.-	€1534.-
Linesmen	€1534.-	€767.-

Table 7.1: Fees per game supervised paid to German soccer referees[45]

Nevertheless, even these high fees evidently did not prevent Robert Hoyzer to enter dubious agreements and to manipulate professional soccer games in Germany. From this follows that merely raising the fees paid to referees per game supervised does not seem to be sufficient to efficiently prevent referees from corruption.

Long-Term Working Contract   Instead, sports associations should think about offering referees a long-term working contract providing for a fixed annual salary, rather than paying referees on a per game basis. A fraction of this annual salary should, however, still be performance-based to retain the incentive effects from the competition between referees.[46]

At the same time, such a working contract should include explicit terms for a breach of contract. These should not only specifically outline the be-

---

[44]   DHB (2009b, p. 8)
[45]   Koch and Maennig (2007, p. 55)
[46]   Part of the annual salary could, for instance, depend on the value of the referee, as determined in Chapter 4.

haviors defined under the term *"sports corruption"* but also list a suitable catalogue of penalties applicable in case of breach.

The advantage of such a long term working contract would be that sports associations could still enforce contractual penalties upon corrupt referees through civil law procedures, even if they exit the sports association and thereby technically escape the power of its authority.

"Referee Pension Fund" In addition to that, the sports association may want to think about creating a *"Referee Pension Fund"*, into which the sports association directly pays a certain percentage of the referee's annual salary that the referee only receives after a corruption free career.[47] In this way, a corrupt referee not only risks to lose his (potential) future income but also the income that he has already earned, which should have a much greater deterring effect.

Such a fund combined with a long term working contract could significantly raise a referee's expected cost k_r from corruption. Yet, such policies only target the referee's incentives for corruption so that they are only helpful for the persecution of Option (2).

7.3   The "Draft Bill"

Finally, re-consider the Draft Bill prepared by the Free State of Bavaria. The implementation of the Draft Bill would in essence have three effects.

First of all, the embedding of sports corruption as an offense under criminal law has a signalling deterrence effect. Especially, given the presently uncertain treatment of sports corruption, as discussed in Chapter 5, the Draft Bill provides for more legal clarity signalling an increased dedication to punish corrupt activities also in the context of sports contests.

Second, by making investigation methods, such as search, attachment,

---

[47]   Maennig (2005, pp. 212 et seq.)

remand, wire-tapping etc., available, which are currently only applicable in the scope of criminal procedure, the Draft Bill would significantly increase the probability of detection for sports defrauders.

Finally, the penalty in case of conviction that wrongdoers risk incurring would substantially increase. There are two reasons. First, the imposition of a punishment of up to ten years of imprisonment for a (severe) offense of sports corruption by far exceeds the severity of the penalties currently applicable by sports associations such as fines, the withdrawal of license, or the exclusion from the sports association, for example. Second, unlike penalties imposed by sports associations, statutory punishments extend well beyond the sports society because they constitute an entry in the criminal record for offenders. Because such a criminal record among others impedes the ease of travelling as well as the prospective job search for offenders, statutory punishments have a much more far reaching impeding influence on the social lives of offenders than the catalogue of penalties available to sports associations.

However, the problem with imprisonment as a punishment for sports corruption is that it creates a high social cost.[48] Thus, it needs to be tested whether an efficient use of fines might be preferable.

Nevertheless, the Draft Bill is an effective anti-corruption policy, as it substantially increases k_r and k_c, which will tremendously lower not only the referee's but also the contestant's incentives to manipulate a game. Therefore, by combining forces with the state in the way as suggested by the Draft Bill would also greatly simplify the persecution of Option (1) for sports associations.

---

[48]   Becker (1968, pp. 179 et seq.), Ehrlich (1996, pp. 63 et seq.)

Chapter 8

Conclusion

In their responsibility to neutrally supervise contests and enforce their underlying rules, sports referees are not a debut of modern times. The so-called *"Hellanodicae"* already took on comparable responsibilities in the supervision and administration of the ancient Olympic Games. Nevertheless, the economic study of referees is still in its initial stage. Up until now, merely empirical and experimental analyses studying the possibility of a referee home bias are available. This dissertation, in its prime part, thereby made the first attempt to explain the rationale behind the institution of sports referees as well as behind the potential threat that corrupt referees pose to the sports industry from a purely economic perspective.

8.1   Summary of Results

Foremost, the analysis showed that the referee must be viewed as an integral part of contests, as he can significantly influence the equilibrium strategies of the contestants. This inference not only applies to sports contests but also to any other type of contest involving a third party supervisor enforcing a pre-specified set of rules. For sports associations, referees thereby become an important instrument of design to uphold the quality of contests and thus to maintain the fan interest in its sports events.

However, the referee's performance is essential for this purpose. Only if the referee's type 1 and type 2 errors are sufficiently low, given the specified penalties awardable subject to the rules of the game, the use of a referee becomes valuable to a sports association. In this regard, it appears to be irrelevant for the value of an honest referee whether he supervises a game between contestants of equal or unequal strength, as long as the disparity between the contestants' characteristics does not affect the referee's errors.

Exploiting his power of discretion, a biased referee, however, necessarily has to raise his type 1 and type 2 errors by a conscious component above his unconscious errors. This unambiguously lowers the game quality, not only relative to a game with an honest referee, but possibly also even relative to a game without any referee at all. In other words, the problem of sports corruption utterly challenges the utility of referees. Thus, it is of eminent importance for the continuing fan interest in sports events that sports leagues proactively establish anti-corruption policies.

For a cost efficient implementation of such anti-corruption policies the prerequisite of a bilateral agreement for sports corruption turns out to play an important role. In particular, it provides sports associations with two options to prevent sports corruption, namely to target either the contestant (Option (1)) or the referee (Option (2)).

From this follows that a sports association must understand not only a referee's but also a contestant's incentives to pursue corrupt endeavors. Above all, sports associations must recognize the interdependence of the contestant's and the referee's incentives for sports corruption originating from the congruent probability of success of a corrupt deal. The derivation of the optimal prize awarded to the winner of a contest and the optimal incentive contract offered to the referee by the sports association illustrates this point.

Anti-Corruption Policy	Option (1)	Option (2)
Responsible Media Coverage	✓	✓
Independent Investigating Institution	✓	✓
Code of Conduct	✓	✓
System of Checks and Balances	✓	✓
Whistleblowing System	✓	✓
Anonymous Help Organization		✓
Technological Assistance		✓
Skill Enhancement of Referees	✓	✓
Efficient Performance Evaluation		✓
Job Rotation:		
(i) Among Performance Evaluators		✓
(ii) Among Referees	✓	
Number of Referees:		
(i) Number of Eligible Referees		✓
(ii) Number of Referees per Game	✓	✓
Secrecy of Referee Assignment	✓	
Referee's Remuneration:		
(i) Long Term Contract		✓
(ii) Referee Pension Fund		✓
Draft Bill	✓	✓

Table 8.1: Anti-corruption policies and their primary usefulness

The results for the optimal incentivization mechanism also reveal that the most cost efficient way to prevent sports corruption is (if feasible) always to already deter the contestant from bribing the referee, rather than only deterring the referee from accepting the contestant's bribe. Thus, anti-corruption mechanisms annihilating the contestant's (i.e. the briber's) incentives for sports corruption would be preferable to those that target the referee's (i.e. the receiver's) incentives. Table 8.1 summarizes the primary usefulness of the anti-corruption policies discussed in Chapter 7 with respect to the pursuit of the two options.

However, so far most sports associations have played down the severity of the problem of sports corruption in Germany. Considering the occasionally extensive financial interests at stake, it might actually be fair to suspect that some sports associations have done so with conditional intent. Either way, the persistent negligence of sports corruption has now led to the ever-more appearing helplessness of sports associations in the fight against sports corruption.

The discussion on the legal treatment of sports corruption in Germany showed that the current regulations of the German Criminal Code are not of much assistance in this regard either, as they generally leave sports defrauders unpunished (as long as the delinquencies are not related to betting fraud). On the one hand, the provisions of the Criminal Code penalizing corruption and corruptibility as such are, due to the evidently unfulfilled yet required elements of offense, simply not applicable. On the other hand, in the scope of crimes against property that are relevant under criminal law, such as fraud and embezzlement, it is in most cases problematic to prove a direct pecuniary damage.

It follows that presently, the prevention of sports corruption in Germany rests solely on the shoulders of autonomous sports associations, which makes the implementation of efficient anti-corruption policies ever more important. The Basketball Bundesliga, already availing of a range of policies targeting the problem of sports corruption, seems to have understood the need for such policies better than most other German sports associations.

However, all sports associations are currently not only limited with respect to their scale of penalties, as they lose their applicable sanctionary authority as soon as sports defrauders exit the sports association, but they also do not have access to efficient instruments for investigating irregularities, making the discovery of corruption cases exceedingly difficult.

Therefore, the Draft Bill put forth by the judiciary of the Free State of Bavaria, asking for a legally protected interest in sports, by making sports fraud a criminal offense, indeed suggests viable and possibly even inevitable means of effectively preventing sports corruption in Germany. However, criminal law is only meant to protect the most important public interests. Thus, the question of whether or not the Draft Bill will be implemented perforcely comes down to the question of whether the sports ethos is in itself a public interest worth protecting. Yet, the wide ranging social influence of all kinds of sports along with the positive social values sports teach in all aspects of our social lives evidently plead for the implementation of the Draft Bill.

But up until the day of the adoption of the Draft Bill, sports associations will have to rely solely on their own feasible anti-corruption policies. Due to the currently ineffective scale of penalties as well as the lack of efficient investigation methods, sports associations would be well advised to also consider positive anti-corruption policies, such as those raising the opportunity cost of sports corruption, aside from negative anti-corruption policies, such as a more severe penalty in case of conviction. Thus, despite their suboptimal cost efficiency (because they target the referee) policies such as an anonymous help organization, long term working contracts for referees as well as a "Referee Pension Fund" should thoroughly be considered.

8.2 Limitations and Proposals for Prospective Research

Of course, this analysis being the first theoretical study on referees still exhibits some limitations. One of them is certainly the implicity of the results on the role of sports referees in Chapter 4, which made their interpretation a bit more difficult than maybe necessary. Therefore, an approach that allows the derivation of explicit results would greatly enhance the technical ease of modelling referees.

Furthermore, the underlying assumption of complete information in Chapter 4 might in some aspects deviate from reality. However, it should also be possible to use the provided framework to discuss a (Bayesian) game of incomplete information. Thus, integrating an additional probability distribution describing the referee's state of bias might be an interesting feature for future research on the impact of biased referees on contests. However, again the implicity of the results might make this analysis exceedingly convoluted.

Also, because the way the game quality is defined is a fundamental element of the analysis in Chapter 4, it might also be interesting to see how varied definitions of the game quality (as defined by sports associations) affect the utility of referees in sports contests.

Another aspect that was left unattended is the effect of varying degrees of risk aversion on the results derived above. Throughout the whole discussion all parties were assumed to be risk neutral. It would, however, be nice to see whether and, if so, how the value of a referee would change when the contestants are risk averse, for example. Intuitively, it would be expected that under the assumption of risk aversion the value of a referee would increase because risk averse individuals tend to be more responsive to imposed controls. For the same reason, it would be expected that the deterrence of sports corruption would be easier with risk averse than with risk neutral (or risk loving) contestants and/or referees.

In any case, an increased attention by economic scholars to the study of sports referees and the problem of sports corruption in the future would be gratifying. The instruments available to economic research certainly have the potential to help obliterating the basis of the Devil's optimism in the soccer match between Heaven and Hell (introducing this thesis). This dissertation hopefully provides some impulses for this purpose.

Appendix

A.1 The Asymmetry Level Maximizing Effort Choices

Contestant j's effort levels in an asymmetric game without a referee are maximized when $\frac{\partial \mu^*_{j_{nr}}}{\partial \alpha} = \frac{\partial s^*_{ji_{nr}}}{\partial \alpha} = 0$. Thus, taking the first order conditions of equations (4.19) and (4.20) yields:

$$\frac{\partial \mu^*_{j_{nr}}}{\partial \alpha} = \frac{t^\mu \frac{\partial g^{nr}[\cdot]}{\partial \alpha}(1+\alpha)V}{c} + \frac{t^\mu g^{nr}[\cdot]V}{c} = 0 \qquad (A.1)$$

$$\frac{\partial s^*_{ji_{nr}}}{\partial \alpha} = \frac{t^s \frac{\partial g^{nr}[\cdot]}{\partial \alpha}(1+\alpha)V}{k} + \frac{t^s g^{nr}[\cdot]V}{k} = 0 \qquad (A.2)$$

Equations (A.1) and (A.2) can be re-arranged to

$$\frac{t^\mu V}{c}\left[\frac{\partial g^{nr}[\cdot]}{\partial \alpha}(1+\alpha) + g^{nr}[\cdot]\right] = 0$$

and

$$\frac{t^s V}{k}\left[\frac{\partial g^{nr}[\cdot]}{\partial \alpha}(1+\alpha) + g^{nr}[\cdot]\right] = 0$$

respectively, which both yield

$$\frac{\partial g^{nr}\left[\cdot\right]}{\partial \alpha}\left(1+\alpha\right) = -g^{nr}\left[\cdot\right]$$

and thus

$$1+\alpha = -\frac{g^{nr}\left[\cdot\right]}{\frac{\partial g^{nr}\left[\cdot\right]}{\partial \alpha}},$$

where $\frac{\partial g^{nr}\left[\cdot\right]}{\partial \alpha} < 0$. Thus, contestant j maximizes his productive effort and his sabotage effort at the asymmetry level $1 + \alpha = -\frac{g^{nr}\left[\cdot\right]}{\frac{\partial g^{nr}\left[\cdot\right]}{\partial \alpha}}$ if contestants value the winner's prize asymmetrically such that $V_i = (1 - \alpha)\,V$ and $V_j = (1+\alpha)\,V$.

A similar result is obtained when contestants are asymmetric in productive playing abilities, i.e. $t_i^\mu = (1-\beta)\,t^\mu$ and $t_j^\mu = (1+\beta)\,t^\mu$. In this case, contestant j will maximize his productive playing effort when $1 + \beta = -\frac{g^{nr}\left[\cdot\right]}{\frac{\partial g^{nr}\left[\cdot\right]}{\partial \beta}}$.

A.2 The Effect of an Asymmetry in Productive Playing Ability on the Referee's Value

The asymmetry in productive playing talents $t_i^\mu = (1 - \beta) t^\mu$ and $t_j^\mu = (1 + \beta) t^\mu$ converts equations (4.10) and (4.11) to

$$\mu_{i_{hr}}^* = \frac{(1 - \beta)t^\mu \left[1 - m(1 + F)\right] g^{hr} \left[\cdot\right] V}{c} \tag{A.3}$$

$$s_{ij_{hr}}^* = \frac{t^s \left[1 - (1 - \overline{m})F\right] g^{hr} \left[\cdot\right] V}{k} \tag{A.4}$$

$$\mu_{j_{hr}}^* = \frac{(1 + \beta)t^\mu \left[1 - m(1 + F)\right] g^{hr} \left[\cdot\right] V}{c} \tag{A.5}$$

$$s_{ji_{hr}}^* = \frac{t^s \left[1 - (1 - \overline{m})F\right] g^{hr} \left[\cdot\right] V}{k} \tag{A.6}$$

for contestant i and j respectively. This leads to the following average game quality

$$\phi_{hr}^a = \frac{\frac{\mu_{i_{hr}}^*}{s_{ji_{hr}}^*} + \frac{\mu_{j_{hr}}^*}{s_{ij_{hr}}^*}}{2} = \frac{\frac{(1-\beta)t^\mu[1-m(1+F)]k}{t^s[1-(1-\overline{m})F]c} + \frac{(1+\beta)t^\mu[1-m(1+F)]k}{t^s[1-(1-\overline{m})F]c}}{2}$$

$$= \frac{1}{2}\frac{t^\mu k}{t^s c}\left[(1 - \beta)\frac{[1 - m(1 + F)]}{[1 - (1 - \overline{m})F]} + (1 + \beta)\frac{[1 - m(1 + F)]}{[1 - (1 - \overline{m})F]}\right] \tag{A.7}$$

which can be re-written as:

$$\phi_{hr}^a = \frac{t^\mu k}{t^s c}\frac{[1 - m(1 + F)]}{[1 - (1 - \overline{m})F]}$$

The value of a referee in a game with asymmetric productive playing talents is therefore given by

$$\frac{\phi_{hr}^a}{\phi_{nr}^a} - 1 = \frac{\frac{t^\mu k}{t^s c} \frac{[1-m(1+F)]}{[1-(1-\overline{m})F]}}{\frac{t^\mu k}{t^s c}} - 1 = \frac{[1 - m(1 + F)]}{[1 - (1 - \overline{m})F]} - 1. \qquad (A.8)$$

Thus, the value of an honest referee is unaffected by an asymmetry in productive playing talents.

A.3 The Effect of a Biased Referee on the Game Quality

From expression (A.7) in Appendix A.2 we already know that the game quality of a game with an honest referee and where $t_i^\mu = (1 - \beta)\, t^\mu$ and $t_j^\mu = (1 + \beta)\, t^\mu$ is given by:

$$\phi_{hr}^a = \frac{1}{2}\frac{t^\mu k}{t^s c}\left[(1 - \beta)\frac{[1 - m(1 + F)]}{[1 - (1 - \overline{m})F]} + (1 + \beta)\frac{[1 - m(1 + F)]}{[1 - (1 - \overline{m})F]}\right]$$

Expression (4.48) provides the game quality of a game with a biased referee and the same asymmetry in productive playing abilities:

$$\phi_{br}^a = \frac{1}{2}\frac{t^\mu k}{t^s c}\left[(1 - \beta)\frac{[1 - m(1 + F)]}{[1 - (1 - \overline{m})F]} + (1 + \beta)\frac{[1 - m_j^b(1 + F)]}{[1 - (1 - \overline{m}_j^b)F]}\right]$$

Thus, in order to show that $\phi_{hr}^a > \phi_{br}^a$, re-write $\phi_{hr}^a > \phi_{br}^a$ as:

$$\frac{1}{2}\frac{t^\mu k}{t^s c}\left[(1 - \beta)\frac{[1 - m(1 + F)]}{[1 - (1 - \overline{m})F]} + (1 + \beta)\frac{[1 - m(1 + F)]}{[1 - (1 - \overline{m})F]}\right] >$$

$$\frac{1}{2}\frac{t^\mu k}{t^s c}\left[(1 - \beta)\frac{[1 - m(1 + F)]}{[1 - (1 - \overline{m})F]} + (1 + \beta)\frac{[1 - m_j^b(1 + F)]}{[1 - (1 - \overline{m}_j^b)F]}\right] \quad (A.9)$$

Rearranging yields

$$\frac{[1 - m(1 + F)]}{[1 - (1 - \overline{m})F]} > \frac{[1 - m_j^b(1 + F)]}{[1 - (1 - \overline{m}_j^b)F]}, \quad (A.10)$$

which, as we know from expression (4.13), always holds because $m < m_j^b$ and $\overline{m} < \overline{m}_j^b$. Hence, it must be true that $\phi_{hr}^a > \phi_{br}^a$, i.e. relative to an honest referee a biased referee reduces the game quality.

A.4 Option (1) - Targeting the Contestant

Pursuing Option (1), the sports association's optimization problem is to

$$\min_{S,V} S + V$$

subject to the constraints

$$E\left[\pi_c\left(q^L, m\right)\right] \leq E\left[\pi_c\left(q^H, m\right)\right] \tag{A.11}$$

$$\max\left\{E\left[\pi_c\left(q^L, m^b\right)\right], E\left[\pi_c\left(q^H, m^b\right)\right]\right\} \leq E\left[\pi_c\left(q^H, m\right)\right] \tag{A.12}$$

$$0 \leq E\left[\pi_r\left(q^H, m\right)\right] \tag{A.13}$$

Condition (A.12) implies that there are two cases that need to be considered:

- Case 1: $\max\left\{E\left[\pi_c\left(q^L, m^b\right)\right], E\left[\pi_c\left(q^H, m^b\right)\right]\right\} = E\left[\pi_c\left(q^L, m^b\right)\right]$

- Case 2: $\max\left\{E\left[\pi_c\left(q^L, m^b\right)\right], E\left[\pi_c\left(q^H, m^b\right)\right]\right\} = E\left[\pi_c\left(q^H, m^b\right)\right]$

After deriving the set of possible solutions for each of the two cases, the ultimate optimal solution of this optimization problem can be determined via comparison of all possibilities subject to their necessary constraints.

Case 1

If $\max \left\{ E\left[\pi_c\left(q^L, m^b\right)\right], E\left[\pi_c\left(q^H, m^b\right)\right]\right\} = E\left[\pi_c\left(q^L, m^b\right)\right]$, the Lagrange function of the sports association's optimization problem is given by:

$$
\begin{aligned}
\mathcal{L} = \ & -(S+V) + \lambda_1 \left(G\left(q^H, m\right) V - c_c\left(q^H\right) - G\left(q^L, m\right) V + c_c\left(q^L\right)\right) \\
& + \lambda_2 (G\left(q^H, m\right) V - c_c\left(q^H\right) - G\left(q^L, m^b\right)\left(V - B\left(m^b\right)\right) \\
& + c_c\left(q^L\right) + k_c\left(m^b\right)) \\
& + \lambda_3 \left(\left(1 + p_r H\left(v_r^L\right) + (1 - p_r) H\left(v_r^H\right)\right) S - c_r\right)
\end{aligned}
$$

Accordingly, the Kuhn-Tucker-Conditions are:

$$
\begin{aligned}
\frac{\partial \mathcal{L}}{\partial V} &= -1 + \lambda_1 \left(G\left(q^H, m\right) - G\left(q^L, m\right)\right) + \lambda_2 \left(G\left(q^H, m\right) - G\left(q^L, m^b\right)\right) \\
&= 0
\end{aligned}
$$

$$
\frac{\partial \mathcal{L}}{\partial S} = -1 + \lambda_3 \left(1 + p_r H\left(v_r^L\right) + (1 - p_r) H\left(v_r^H\right)\right) = 0
$$

$$
\begin{aligned}
\lambda_1 &\geq 0 \ \left(= 0 \text{ if } E\left[\pi_c\left(q^L, m\right)\right] < E\left[\pi_c\left(q^H, m\right)\right]\right) \\
\lambda_2 &\geq 0 \ \left(= 0 \text{ if } E\left[\pi_c\left(q^L, m^b\right)\right] < E\left[\pi_c\left(q^H, m\right)\right]\right) \\
\lambda_3 &\geq 0 \ \left(= 0 \text{ if } 0 < E\left[\pi_r\left(q^H, m\right)\right]\right)
\end{aligned}
$$

Optimal V:

Considering the first Kuhn-Tucker-Condition it becomes evident that for $\frac{\partial \mathcal{L}}{\partial V} = 0$ to be satisfiable it must hold that $\lambda_1 > 0$ because $G\left(q^H, m\right) - G\left(q^L, m\right) > 0$ and $G\left(q^H, m\right) - G\left(q^L, m^b\right) < 0$. Nevertheless, there are still two possibilities that need to be considered here, namely (i) $\lambda_1 > 0$ and $\lambda_2 = 0$, and (ii) $\lambda_1 > 0$ and $\lambda_2 > 0$.

(i) $\lambda_1 > 0$ and $\lambda_2 = 0$: If $\lambda_1 > 0$ and $\lambda_2 = 0$, condition (A.11) must be binding and determines the optimal winner's prize V:

$$V = \frac{c_c\left(q^H\right) - c_c\left(q^L\right)}{G\left(q^H, m\right) - G\left(q^L, m\right)} \tag{A.14}$$

Plugging expression (A.14) into condition (A.12), given by $E\left[\pi_c\left(q^L, m^b\right)\right] \leq E\left[\pi_c\left(q^H, m\right)\right]$ in Case 1, then requires that

$$\frac{c_c\left(q^H\right) - c_c\left(q^L\right)}{G\left(q^H, m\right) - G\left(q^L, m\right)} \leq \frac{G\left(q^L, m^b\right) B\left(m^b\right) + k_c\left(m^b\right) + c_c\left(q^L\right) - c_c\left(q^H\right)}{G\left(q^L, m^b\right) - G\left(q^H, m\right)} \tag{A.15}$$

holds.

(ii) $\lambda_1 > 0$ and $\lambda_2 > 0$: If $\lambda_1 > 0$ and $\lambda_2 > 0$, not only condition (A.11) but also condition (A.12) would have to be binding. Accordingly, the optimal winner's prize would be determined by:

$$\begin{aligned} V &= \frac{c_c\left(q^H\right) - c_c\left(q^L\right)}{G\left(q^H, m\right) - G\left(q^L, m\right)} \\[2mm] &= \frac{G\left(q^L, m^b\right) B\left(m^b\right) + k_c\left(m^b\right) + c_c\left(q^L\right) - c_c\left(q^H\right)}{G\left(q^L, m^b\right) - G\left(q^H, m\right)} \end{aligned} \tag{A.16}$$

Optimal S:

It is readily observed that $\lambda_3 > 0$ must hold for the second Kuhn-Tucker-Condition $\frac{\partial \mathcal{L}}{\partial S} = 0$ to be satisfiable because $1 + p_r H\left(v_r^L\right) + (1 - p_r) H\left(v_r^H\right) > 0$. From this follows that condition (A.13) (i.e. the referee's participation constraint for honestly supervising a game), is binding and determines the optimal S:

$$S = \frac{c_r}{1 + p_r H\left(v_r^L\right) + (1 - p_r) H\left(v_r^H\right)} \tag{A.17}$$

Case 2

If $\max \left\{ E\left[\pi_c \left(q^L, m^b \right) \right], E\left[\pi_c \left(q^H, m^b \right) \right] \right\} = E\left[\pi_c \left(q^H, m^b \right) \right]$, the Lagrange
function of the association's optimization problem is given by:

$$
\begin{aligned}
\mathcal{L} = {} & -(S+V) + \lambda_1 \left(G\left(q^H, m \right) V - c_c \left(q^H \right) - G\left(q^L, m \right) V + c_c \left(q^L \right) \right) \\
& + \lambda_2 (G\left(q^H, m \right) V - c_c \left(q^H \right) - G\left(q^H, m^b \right) \left(V - B\left(m^b \right) \right) \\
& + c_c \left(q^H \right) + k_c \left(m^b \right)) \\
& + \lambda_3 \left(\left(1 + p_r H\left(v_r^L \right) + (1 - p_r) H\left(v_r^H \right) \right) S - c_r \right)
\end{aligned}
$$

Accordingly, the Kuhn-Tucker-Conditions are:

$$
\begin{aligned}
\frac{\partial \mathcal{L}}{\partial V} &= -1 + \lambda_1 \left(G\left(q^H, m \right) - G\left(q^L, m \right) \right) + \lambda_2 \left(G\left(q^H, m \right) - G\left(q^H, m^b \right) \right) \\
&= 0
\end{aligned}
$$

$$
\frac{\partial \mathcal{L}}{\partial S} = -1 + \lambda_3 \left(1 + p_r H\left(v_r^L \right) + (1 - p_r) H\left(v_r^H \right) \right) = 0
$$

$$
\begin{aligned}
\lambda_1 &\geq 0 \ \left(= 0 \text{ if } E\left[\pi_c \left(q^L, m \right) \right] < E\left[\pi_c \left(q^H, m \right) \right] \right) \\
\lambda_2 &\geq 0 \ \left(= 0 \text{ if } E\left[\pi_c \left(q^H, m^b \right) \right] < E\left[\pi_c \left(q^H, m \right) \right] \right) \\
\lambda_3 &\geq 0 \ \left(= 0 \text{ if } 0 < E\left[\pi_r \left(q^H, m \right) \right] \right)
\end{aligned}
$$

Optimal V:

Again, for $\frac{\partial \mathcal{L}}{\partial V} = 0$ to be satisfiable, it must be that $\lambda_1 > 0$ because
$G\left(q^H, m \right) - G\left(q^L, m \right) > 0$ and $G\left(q^H, m \right) - G\left(q^H, m^b \right) < 0$. Yet, again
the two remaining possibilities, (i) $\lambda_1 > 0$ and $\lambda_2 = 0$ and (ii) $\lambda_1 > 0$ and
$\lambda_2 > 0$, still have to be considered here.

(i) $\lambda_1 > 0$ **and** $\lambda_2 = 0$: If $\lambda_1 > 0$ and $\lambda_2 = 0$, condition (A.11) is once more binding and determines the optimal winner's prize V:

$$V = \frac{c_c\left(q^H\right) - c_c\left(q^L\right)}{G\left(q^H, m\right) - G\left(q^L, m\right)}$$

Plugging this into condition (A.12), given by $E\left[\pi_c\left(q^H, m^b\right)\right] \leq E\left[\pi_c\left(q^H, m\right)\right]$ in Case 2, then requires that:

$$\frac{c_c\left(q^H\right) - c_c\left(q^L\right)}{G\left(q^H, m\right) - G\left(q^L, m\right)} \leq \frac{G\left(q^H, m^b\right) B\left(m^b\right) + k_c\left(m^b\right)}{G\left(q^H, m^b\right) - G\left(q^H, m\right)} \tag{A.18}$$

(ii) $\lambda_1 > 0$ **and** $\lambda_2 > 0$: If $\lambda_1 > 0$ and $\lambda_2 > 0$, not only condition (A.11) but also condition (A.12) would have to be binding. Therefore, the optimal winner's prize would be determined by:

$$V = \frac{c_c\left(q^H\right) - c_c\left(q^L\right)}{G\left(q^H, m\right) - G\left(q^L, m\right)} = \frac{G\left(q^H, m^b\right) B\left(m^b\right) + k_c\left(m^b\right)}{G\left(q^H, m^b\right) - G\left(q^H, m\right)} \tag{A.19}$$

Optimal S:

The optimal incentive contract S in Case 2 is identical to that in Case 1.

Case 1 versus Case 2

In both cases, it appears that the possibility where $\lambda_1 > 0$ and $\lambda_2 > 0$ does not provide an inherently new solution. By necessity, Case 1(i) entails the solution of Case 1(ii), and Case 2(i) entails the solution of Case 2(ii). Thus, the selection of possible optimal solutions reduces to Case 1(i) and Case 2(i).

Now, note that both Case 1(i) and Case 2(i) suggest that equation (A.14) determines the level of the optimal winner's prize V. At this winner's prize, it turns out, however, that $E\left[\pi_c\left(q^L, m^b\right)\right] > E\left[\pi_c\left(q^H, m^b\right)\right]$, which suggests that Case 1(i) must provide the optimal solution. This is shown below.

Proof. Evidently, $E\left[\pi_c\left(q^L, m^b\right)\right] > E\left[\pi_c\left(q^H, m^b\right)\right]$ holds as long as

$$\frac{c_c\left(q^H\right) - c_c\left(q^L\right)}{G\left(q^H, m^b\right) - G\left(q^L, m^b\right)} + B\left(m^b\right) > V. \tag{A.20}$$

Plugging equation (A.14) into condition (A.20) gives:

$$\frac{c_c\left(q^H\right) - c_c\left(q^L\right)}{G\left(q^H, m^b\right) - G\left(q^L, m^b\right)} + B\left(m^b\right) > \frac{c_c\left(q^H\right) - c_c\left(q^L\right)}{G\left(q^H, m\right) - G\left(q^L, m\right)} \tag{A.21}$$

Unambiguously, condition (A.21) holds for any bribe $B\left(m^b\right) \geq 0$, as $G\left(q^H, m^b\right) - G\left(q^L, m^b\right) < G\left(q^H, m\right) - G\left(q^L, m\right)$. It follows that indeed at the optimal winner's prize (A.14) $E\left[\pi_c\left(q^L, m^b\right)\right] > E\left[\pi_c\left(q^H, m^b\right)\right]$ holds. ∎

Hence, the optimal solution for the winner's prize V under Option (1) is provided by Case 1(i), while the optimal incentive contract S (is independent of the case) given by equation (A.17).

A.5 Option (2) - Targeting the Referee

Pursuing Option (2), the sports association's optimization problem is again to

$$\min_{S,V} S + V$$

but subject to a different set of constraints. The necessary constraints to be satisfied here are given by:

$$E\left[\pi_c\left(q^L, m\right)\right] \leq E\left[\pi_c\left(q^H, m\right)\right] \tag{A.22}$$

$$E\left[\pi_c\left(q^H, m\right)\right] \leq \max\left\{E\left[\pi_c\left(q^L, m^b\right)\right], E\left[\pi_c\left(q^H, m^b\right)\right]\right\} \tag{A.23}$$

$$E\left[\pi_r\left(q^L, m^b\right)\right] \leq E\left[\pi_r\left(q^H, m\right)\right] \text{ (in Case 1)} \tag{A.24}$$

$$E\left[\pi_r\left(q^H, m^b\right)\right] \leq E\left[\pi_r\left(q^H, m\right)\right] \text{ (in Case 2)} \tag{A.25}$$

Due to the nature of condition (A.23), the same two cases, as discussed in Option (1), need to be considered in Option (2) as well. However, now the contestant has to be willing to bribe the referee. Hence, corruption prevention requires the sports association to deter the referee from entering a side contract, where the necessary condition to do so differs across the two cases. This is because the referee can only attain $E\left[\pi_r\left(q^L, m^b\right)\right]$ in Case 1, while he could attain $E\left[\pi_r\left(q^H, m^b\right)\right]$ $\left(> E\left[\pi_r\left(q^L, m^b\right)\right]\right)$ in Case 2.

Case 1

If $\max\left\{E\left[\pi_c\left(q^L,m^b\right)\right],E\left[\pi_c\left(q^H,m^b\right)\right]\right\}=E\left[\pi_c\left(q^L,m^b\right)\right]$, the Lagrange function of the sports association's optimization problem is given by:

$$
\begin{aligned}
\mathcal{L} = &-\left(S+V\right)+\lambda_1\left(G\left(q^H,m\right)V-c_c\left(q^H\right)-G\left(q^L,m\right)V+c_c\left(q^L\right)\right)\\
&+\lambda_2(G\left(q^L,m^b\right)\left(V-B\left(m^b\right)\right)-c_c\left(q^L\right)-k_c\left(m^b\right)\\
&-G\left(q^H,m\right)V+c_c\left(q^H\right))\\
&+\lambda_3((1+p_r H\left(v_r^L\right)+(1-p_r)H\left(v_r^H\right))S-c_r\\
&-\left(1+H\left(v_r^L\right)\right)S-G\left(q^L,m^b\right)B\left(m^b\right)+c_r+k_r\left(m^b\right))
\end{aligned}
$$

Accordingly, the Kuhn-Tucker-Conditions are:

$$
\begin{aligned}
\frac{\partial\mathcal{L}}{\partial V} &= -1+\lambda_1\left(G\left(q^H,m\right)-G\left(q^L,m\right)\right)+\lambda_2\left(G\left(q^L,m^b\right)-G\left(q^H,m\right)\right)\\
&= 0
\end{aligned}
$$

$$
\frac{\partial\mathcal{L}}{\partial S} = -1+\lambda_3\left(1-p_r\right)\left(H\left(v_r^H\right)-H\left(v_r^L\right)\right)=0
$$

$$
\begin{aligned}
\lambda_1 &\geq 0\ \left(=0\text{ if }E\left[\pi_c\left(q^L,m\right)\right]<E\left[\pi_c\left(q^H,m\right)\right]\right)\\
\lambda_2 &\geq 0\ \left(=0\text{ if }E\left[\pi_c\left(q^H,m\right)\right]<E\left[\pi_c\left(q^L,m^b\right)\right]\right)\\
\lambda_3 &\geq 0\ \left(=0\text{ if }E\left[\pi_r\left(q^L,m^b\right)\right]<E\left[\pi_r\left(q^H,m\right)\right]\right)
\end{aligned}
$$

Optimal V:

Note that not only $G\left(q^H,m\right)-G\left(q^L,m\right)>0$ but also $G\left(q^L,m^b\right)-G\left(q^H,m\right)>0$. Thus, while at least one of the constraints has to be binding, λ_1 and/or λ_2 could be positive so that it is not clear, which of the constraints determines V. The derivation of the optimal V under Option (2) therefore requires the discussion of three possibilities: (i) $\lambda_1>0$ and $\lambda_2=0$, (ii) $\lambda_2>0$ and $\lambda_1=0$, and (iii) $\lambda_1>0$ and $\lambda_2>0$.

(i) $\lambda_1 > 0$ **and** $\lambda_2 = 0$: If $\lambda_1 > 0$ and $\lambda_2 = 0$, condition (A.22) becomes binding so that equation (A.14) replicated below again determines V.

$$V = \frac{c_c\left(q^H\right) - c_c\left(q^L\right)}{G\left(q^H, m\right) - G\left(q^L, m\right)}$$

Plugging this into condition (A.23), given by $E\left[\pi_c\left(q^H, m\right)\right] \leq E\left[\pi_c\left(q^L, m^b\right)\right]$ in Case 1, now, however, requires that

$$\frac{c_c\left(q^H\right) - c_c\left(q^L\right)}{G\left(q^H, m\right) - G\left(q^L, m\right)} \geq \frac{G\left(q^L, m^b\right) B\left(m^b\right) + k_c\left(m^b\right) + c_c\left(q^L\right) - c_c\left(q^H\right)}{G\left(q^L, m^b\right) - G\left(q^H, m\right)}$$
$$\text{(A.26)}$$

holds.

(ii) $\lambda_2 > 0$ **and** $\lambda_1 = 0$: If $\lambda_2 > 0$ and $\lambda_1 = 0$, $E\left[\pi_c\left(q^H, m\right)\right] \leq E\left[\pi_c\left(q^L, m^b\right)\right]$ becomes binding so that the optimal winner's prize would be determined by:

$$V = \frac{G\left(q^L, m^b\right) B\left(m^b\right) + k_c\left(m^b\right) + c_c\left(q^L\right) - c_c\left(q^H\right)}{G\left(q^L, m^b\right) - G\left(q^H, m\right)} \qquad \text{(A.27)}$$

Plugging equation (A.27) into condition (A.22) would then require for this solution that

$$\frac{c_c\left(q^H\right) - c_c\left(q^L\right)}{G\left(q^H, m\right) - G\left(q^L, m\right)} \leq \frac{G\left(q^L, m^b\right) B\left(m^b\right) + k_c\left(m^b\right) + c_c\left(q^L\right) - c_c\left(q^H\right)}{G\left(q^L, m^b\right) - G\left(q^H, m\right)}$$
$$\text{(A.28)}$$

holds.

(iii) $\lambda_1 > 0$ **and** $\lambda_2 > 0$: If $\lambda_1 > 0$ and $\lambda_2 > 0$, both $E\left[\pi_c\left(q^L, m\right)\right] \leq E\left[\pi_c\left(q^H, m\right)\right]$ and $E\left[\pi_c\left(q^H, m\right)\right] \leq E\left[\pi_c\left(q^L, m^b\right)\right]$ are binding so that the optimal winner's prize would be given by:

$$V = \frac{c_c\left(q^H\right) - c_c\left(q^L\right)}{G\left(q^H, m\right) - G\left(q^L, m\right)}$$

$$= \frac{G\left(q^L, m^b\right) B\left(m^b\right) + k_c\left(m^b\right) + c_c\left(q^L\right) - c_c\left(q^H\right)}{G\left(q^L, m^b\right) - G\left(q^H, m\right)} \quad \text{(A.29)}$$

Optimal S:

Because $(1 - p_r)\left(H\left(v_r^H\right) - H\left(v_r^L\right)\right) > 0$, for $\frac{\partial \mathcal{L}}{\partial S} = 0$ to be satisfied, it is clear that $\lambda_3 > 0$ must hold. Accordingly, condition (A.24) must be binding so that the optimal S is determined by:

$$S = \frac{G\left(q^L, m^b\right) B\left(m^b\right) - k_r\left(m^b\right)}{(1 - p_r)\left(H\left(v_r^H\right) - H\left(v_r^L\right)\right)} \quad \text{(A.30)}$$

Case 2

If $\max\left\{E\left[\pi_c\left(q^L, m^b\right)\right], E\left[\pi_c\left(q^H, m^b\right)\right]\right\} = E\left[\pi_c\left(q^H, m^b\right)\right]$, the Lagrange function of the sports association's optimization problem is given by:

$$
\begin{aligned}
\mathcal{L} = & -(S + V) + \lambda_1\left(G\left(q^H, m\right) V - c_c\left(q^H\right) - G\left(q^L, m\right) V + c_c\left(q^L\right)\right) \\
& + \lambda_2(G\left(q^H, m^b\right)\left(V - B\left(m^b\right)\right) - c_c\left(q^H\right) - k_c\left(m^b\right) \\
& - G\left(q^H, m\right) V + c_c\left(q^H\right)) \\
& + \lambda_3((1 + p_r H\left(v_r^L\right) + (1 - p_r) H\left(v_r^H\right)) S - c_r \\
& - \left(1 + H\left(v_r^L\right)\right) S - G\left(q^H, m^b\right) B\left(m^b\right) + c_r + k_r\left(m^b\right))
\end{aligned}
$$

Accordingly, the Kuhn-Tucker-Conditions are:

$$\frac{\partial \mathcal{L}}{\partial V} = -1 + \lambda_1 \left(G\left(q^H, m\right) - G\left(q^L, m\right) \right) + \lambda_2 \left(G\left(q^H, m^b\right) - G\left(q^H, m\right) \right)$$
$$= 0$$

$$\frac{\partial \mathcal{L}}{\partial S} = -1 + \lambda_3 \left(1 - p_r\right) \left(H\left(v_r^H\right) - H\left(v_r^L\right) \right) = 0$$

$$\lambda_1 \geq 0 \ \left(= 0 \text{ if } E\left[\pi_c\left(q^L, m\right)\right] < E\left[\pi_c\left(q^H, m\right)\right] \right)$$
$$\lambda_2 \geq 0 \ \left(= 0 \text{ if } E\left[\pi_c\left(q^H, m\right)\right] < E\left[\pi_c\left(q^H, m^b\right)\right] \right)$$
$$\lambda_3 \geq 0 \ \left(= 0 \text{ if } E\left[\pi_r\left(q^H, m^b\right)\right] < E\left[\pi_r\left(q^H, m\right)\right] \right)$$

Optimal V:

   Again, note that not only $G\left(q^H, m\right) - G\left(q^L, m\right) > 0$ but also $G\left(q^H, m^b\right) - G\left(q^H, m\right) > 0$. Thus, both λ_1 and/or λ_2 could be positive so that it is not clear, which of the constraints determines V. The derivation of the optimal V therefore once more requires the discussion of the three possibilities (i) $\lambda_1 > 0$ and $\lambda_2 = 0$, (ii) $\lambda_2 > 0$ and $\lambda_1 = 0$, and (iii) $\lambda_1 > 0$ and $\lambda_2 > 0$.

(i) $\lambda_1 > 0$ and $\lambda_2 = 0$: If $\lambda_1 > 0$ and $\lambda_2 = 0$, condition (A.22) becomes binding so that equation (A.14) replicated below again determines V.

$$V = \frac{c_c\left(q^H\right) - c_c\left(q^L\right)}{G\left(q^H, m\right) - G\left(q^L, m\right)}$$

Plugging this into condition (A.23), given by $E\left[\pi_c\left(q^H, m\right)\right] \leq E\left[\pi_c\left(q^H, m^b\right)\right]$ in Case 2, would require that

$$\frac{c_c\left(q^H\right) - c_c\left(q^L\right)}{G\left(q^H, m\right) - G\left(q^L, m\right)} \geq \frac{G\left(q^H, m^b\right) B\left(m^b\right) + k_c\left(m^b\right)}{G\left(q^H, m^b\right) - G\left(q^H, m\right)} \tag{A.31}$$

holds.

(ii) $\lambda_2 > 0$ and $\lambda_1 = 0$: If $\lambda_2 > 0$ and $\lambda_1 = 0$, $E\left[\pi_c\left(q^H, m\right)\right] \leq E\left[\pi_c\left(q^H, m^b\right)\right]$ becomes binding so that the optimal winner's prize would be given by:

$$V = \frac{G\left(q^H, m^b\right) B\left(m^b\right) + k_c\left(m^b\right)}{G\left(q^H, m^b\right) - G\left(q^H, m\right)} \tag{A.32}$$

Plugging equation (A.32) into condition (A.22) would then require for this solution that

$$\frac{c_c\left(q^H\right) - c_c\left(q^L\right)}{G\left(q^H, m\right) - G\left(q^L, m\right)} \leq \frac{G\left(q^H, m^b\right) B\left(m^b\right) + k_c\left(m^b\right)}{G\left(q^H, m^b\right) - G\left(q^H, m\right)} \tag{A.33}$$

holds.

(iii) $\lambda_1 > 0$ and $\lambda_2 > 0$: If $\lambda_1 > 0$ and $\lambda_2 > 0$, both $E\left[\pi_c\left(q^L, m\right)\right] \leq E\left[\pi_c\left(q^H, m\right)\right]$ and $E\left[\pi_c\left(q^H, m\right)\right] \leq E\left[\pi_c\left(q^H, m^b\right)\right]$ are binding so that the optimal winner's prize would be given by:

$$V = \frac{c_c\left(q^H\right) - c_c\left(q^L\right)}{G\left(q^H, m\right) - G\left(q^L, m\right)} = \frac{G\left(q^H, m^b\right) B\left(m^b\right) + k_c\left(m^b\right)}{G\left(q^H, m^b\right) - G\left(q^H, m\right)} \tag{A.34}$$

Optimal S:

As in Case 1, because $(1 - p_r)\left(H\left(v_r^H\right) - H\left(v_r^L\right)\right) > 0$, for $\frac{\partial \mathcal{L}}{\partial S} = 0$ to be satisfied, it is clear that $\lambda_3 > 0$ must hold. Accordingly, condition (A.25) must be binding so that the optimal S is determined by:

$$S = \frac{G\left(q^H, m^b\right) B\left(m^b\right) - k_r\left(m^b\right)}{(1 - p_r)\left(H\left(v_r^H\right) - H\left(v_r^L\right)\right)} \tag{A.35}$$

Case 1 versus Case 2

Again, it seems that in both cases the possibility where $\lambda_1 > 0$ and $\lambda_2 > 0$ does not provide an inherently new solution. In fact, by necessity both Case 1(i) and Case 1(ii) entail the solution of Case 1(iii), while both Case 2(i) and Case 2(ii) entail the solution of Case 2(iii). Thus, the selection of possible optimal solutions reduces to Case 1(i) and Case 1(ii) as well as Case 2(i) and Case 2(ii).

Now, recall from condition (A.21) that at the optimal winner's prize (A.14) $\max\left\{E\left[\pi_c\left(q^L, m^b\right)\right], E\left[\pi_c\left(q^H, m^b\right)\right]\right\} = E\left[\pi_c\left(q^L, m^b\right)\right]$, so that condition (A.31) can not be a necessary condition for the winner's prize (A.14). Accordingly, the solution provided by Case 2(i) can not be optimal.

Comparing the solutions of Case 1(ii) and Case 2(ii) in more detail, it in addition becomes evident that the solution of Case 1(ii) dominates the solution of Case 2(ii). This is shown below.

Proof. To demonstrate that the solution of Case 1(ii) dominates the solution of Case 2(ii), re-arrange condition (A.28) and condition (A.33), i.e.

$$\frac{c_c\left(q^H\right) - c_c\left(q^L\right)}{G\left(q^H, m\right) - G\left(q^L, m\right)} \leq \frac{G\left(q^L, m^b\right) B\left(m^b\right) + k_c\left(m^b\right) + c_c\left(q^L\right) - c_c\left(q^H\right)}{G\left(q^L, m^b\right) - G\left(q^H, m\right)}$$

and

$$\frac{c_c\left(q^H\right) - c_c\left(q^L\right)}{G\left(q^H, m\right) - G\left(q^L, m\right)} \leq \frac{G\left(q^H, m^b\right) B\left(m^b\right) + k_c\left(m^b\right)}{G\left(q^H, m^b\right) - G\left(q^H, m\right)}$$

to

$$\frac{G\left(q^L, m^b\right)}{G\left(q^H, m\right) - G\left(q^L, m\right)} - \frac{G\left(q^L, m^b\right) B\left(m^b\right)}{c_c\left(q^H\right) - c_c\left(q^L\right)} + 1 \leq$$

$$\frac{G\left(q^H, m\right)}{G\left(q^H, m\right) - G\left(q^L, m\right)} + \frac{k_c\left(m^b\right)}{c_c\left(q^H\right) - c_c\left(q^L\right)} \tag{A.36}$$

and

$$\frac{G\left(q^H, m^b\right)}{G\left(q^H, m\right) - G\left(q^L, m\right)} - \frac{G\left(q^H, m^b\right) B\left(m^b\right)}{c_c\left(q^H\right) - c_c\left(q^L\right)} \leq$$

$$\frac{G\left(q^H, m\right)}{G\left(q^H, m\right) - G\left(q^L, m\right)} + \frac{k_c\left(m^b\right)}{c_c\left(q^H\right) - c_c\left(q^L\right)} \tag{A.37}$$

respectively. While the RHS of the conditions (A.36) and (A.37) is identical, the LHS of condition (A.36) unambiguously exceeds the LHS of condition (A.37), i.e.

$$\frac{G\left(q^H, m^b\right)}{G\left(q^H, m\right) - G\left(q^L, m\right)} - \frac{G\left(q^H, m^b\right) B\left(m^b\right)}{c_c\left(q^H\right) - c_c\left(q^L\right)} <$$

$$\frac{G\left(q^L, m^b\right)}{G\left(q^H, m\right) - G\left(q^L, m\right)} - \frac{G\left(q^L, m^b\right) B\left(m^b\right)}{c_c\left(q^H\right) - c_c\left(q^L\right)} + 1. \tag{A.38}$$

Condition (A.38) can in turn be re-arranged to:

$$\frac{c_c\left(q^H\right) - c_c\left(q^L\right)}{G\left(q^H, m\right) - G\left(q^L, m\right)} - \frac{c_c\left(q^H\right) - c_c\left(q^L\right)}{G\left(q^H, m^b\right) - G\left(q^L, m^b\right)} < B\left(m^b\right) \tag{A.39}$$

Because $G\left(q^H, m\right) - G\left(q^L, m\right) > G\left(q^H, m^b\right) - G\left(q^L, m^b\right)$,

$$\frac{c_c\left(q^H\right) - c_c\left(q^L\right)}{G\left(q^H, m\right) - G\left(q^L, m\right)} - \frac{c_c\left(q^H\right) - c_c\left(q^L\right)}{G\left(q^H, m^b\right) - G\left(q^L, m^b\right)} < 0$$

so that condition (A.39) is satisfied for any bribe $B\left(m^b\right) \geq 0$.

From this follows that condition (A.28) must be more restrictive than condition (A.33) for any bribe $B\left(m^{b}\right) \geq 0$, which implies that:

$$\frac{G\left(q^{L}, m^{b}\right) B\left(m^{b}\right) + k_{c}\left(m^{b}\right) + c_{c}\left(q^{L}\right) - c_{c}\left(q^{H}\right)}{G\left(q^{L}, m^{b}\right) - G\left(q^{H}, m\right)} <$$

$$\frac{G\left(q^{H}, m^{b}\right) B\left(m^{b}\right) + k_{c}\left(m^{b}\right)}{G\left(q^{H}, m^{b}\right) - G\left(q^{H}, m\right)} \qquad (A.40)$$

As a result, the solution provided by Case 1(ii) dominates the solution provided by Case 2(ii) in terms of cost efficiency. ∎

Finally, observe that the solution in Case 1(ii) of Option (2) is in turn dominated by the solution provided by Case 1(i) of Option (1), where condition (A.28) must also be satisfied. In fact, condition (A.28) itself proves that the optimal winner's prize in Case 1(i) of Option (1) bests the winner's prize in Case 1(ii) of Option (2) in terms of cost efficiency.

Therefore, Case 1(i) describes the optimal solution for the winner's prize V in Option (2). This also accords with the fact that the optimal incentive contract S of Case 1 is clearly more cost efficient than that of Case 2, as $G\left(q^{L}, m^{b}\right) < G\left(q^{H}, m^{b}\right)$.

References

ABBINK K. (2004), "Staff Rotation as an Anti-Corruption Policy: An Experimental Study" in *European Journal of Political Economy 20*, 887-906

AHRENS P. (2005), *Ein Urteil, auf das niemand gewettet hätte*, [online] last access: 20.11.2009; available at: http://www.spiegel.de/sport/fussball/ 0,1518,385504,00.html

AMANN E. AND LEININGER W. (1996), "Asymmetric All-Pay Auctions with Incomplete Information: The Two-Player Case" in *Games and Economic Behavior 14(1)*, 1-18

ANDERSEN J. S. (2006), "Play the Game. Reaktionen einer global operierenden Bewusstseinsindustrie" in Weinrich J. (ed.), *Korruption im Sport: Mafiose Dribblings-Organisiertes Schweigen*, Leipzig, 79-93

ANDVIG J. C. AND FJELDSTAD O. H. (2001), *Corruption - A Review of Contemporary Research*, NUPI Report No. 268, Norwegian Institute of International Affairs, Oslo

ANDVIG J. C. AND MOENE K. O. (1990), "How Corrupt May Corrupt" in *Journal of Economic Behavior and Organization 13*, 63-76

ARBATSKAYA M. AND MIALON H. M. (2008), "Multi-Activity Contests" in *Economic Theory 43(1)*, 23-43

ASHELM M. (2009), "Im Sumpf der Korruption" in *Frankfurter Allgemeine*, 22.11.2009, 17

BAC M. (1996), "Corruption and Supervision Costs in Hierarchies" in *Journal of Comparative Economics 22(2)*, 99-118

BAIK K. (1994), "Effort Levels in Contests with Two Asymmetric Players" in *Southern Economic Journal 61*, 367-378

BARDHAN P. K. (1997), "Corruption and Development: A Review of Issues" in *The Journal of Economic Literature 35*, 1320-1346

BARDHAN P. K. (2006), "The Economist's Approach to the Problem of Corruption" in *World Development 34(2)*, 341-348

BARNEY R. K. (2004), "Prologue: The Ancient Games" in Findling J. E. and Pelle K. D. (eds.), *Encyclopedia of the Modern Olympic Movement*, Westport: Greenwood Press, xxiii-xxxix

BAYE M. R., KOVENOCK D. AND DEVRIES C. G. (1996), "The All-Pay Auction with Complete Information" in *Economic Theory 8*, 291-305

BAYE M. R., KOVENOCK D. AND DEVRIES C. G. (2005), "Comparative Analysis of Litigation Systems: An Auction-Theoretic Approach" in *Economic Journal 115(505)*, 583-601

BECKER G. (1968), "Crime and Punishment: An Economic Approach" in *Journal of Political Economy 76*, 169-217

BESLEY T. AND MCLAREN J. (1993), "Taxes and Bribery: The Role of Wage Incentives" in *Economic Journal 103(416)*, 119-141

BGH (1975), "Decision of February 27, 1975 - 4StR 571/74" in *Neue Juristische Wochenschrift*, 1234-1236

BGH (2007), "Decision of December 15, 2006 - StR 181/06" in *Neue Juristische Wochenschrift*, 782-788

BIERMANN C. AND WULZINGER M. (2008), *Er würde mich umbringen*, [online] last access: 26.01.2011; available at: http://www.spiegel.de/spiegel/print/d-59673705.html

BLISS C. AND DI TELLA R. (1997), "Does Competition Kill Corruption?" in *Journal of Political Economy 105*, 1001-1023

BURAIMO B., FORREST D. AND SIMMONS R. (2010), "The Twelfth Man? Refereeing Bias in English and German Soccer" in *Journal of the Royal Statistical Society 173(2), Series A*, 431-449

CHAND S. K. AND MOENE K. O. (1997), *Controlling Fiscal Corruption*, IMF Working Paper WP/97/100

CHEN K.-P. (2003), "Sabotage in Promotion Tournaments" in *Journal of Law, Economics and Organization 19*, 119-140

CLARK D. J. AND RIIS C. (1998), "Contest Success Functions: An Extension" in *Economic Theory 11*, 201-204

CLARK D. J. AND RIIS C. (2000), "Allocation Efficiency in a Competitive Bribery Game" in *Journal of Economic Behavior and Organization 42(1)*, 109-124

CORNES R. AND HARTLEY R. (2005), "Asymmetric Contests with General Technologies" in *Economic Theory 26(4)*, 923-946

DASGUPTA D. AND STIGLITZ J. (1980), "Uncertainty, Industrial Structure, and the Speed of R&D" in *Bell Journal of Economics*, 1-28

DAWSON P., DOBSON S., GODDARD J. AND WILSON J. (2007), "Are football referees really biased and inconsistent? Evidence from the English Premier League" in *Journal of the Royal Statistical Society 170, Series A*, 231-250

DBB (2008), *Schiedsrichterordnung*, [online] last access: 22.10.2010; available at: http://www.bbsr.de/regel/ord/ord_sr2002.htm

DBB (2009), *Satzung*, [online] last access: 23.01.2011; available at: http://www.basketballbund.de/basketballbund/de/service/formulare_downloads/satzung_und_ordnungen/ 19471.html

DFB (2010a), *Schiedsrichterordnung*, [online] last access: 22.10.2010; available at: http://www.dfb.de/index.php?id=11003

DFB (2010b), *Satzung*, [online] last access: 22.10.2010; available at: http://www.dfb.de/index.php?id=11003

DFB (2010c), *Anti-Doping-Richtlinien*, [online] last access: 25.01.2011; available at: http://www.dfb.de/uploads/media/Anti-Doping-Richtlinien.pdf

DHB (2007), *Schiedsrichterordnung*, [online] last access: 22.10.2010; available at: http://www.dhb.de/index.php?id=42

DHB (2008), *Satzung*, [online] last access: 23.01.2011; available at: http://www.dhb.de/index.php?id=42

DHB (2009a), *Anti-Doping Reglement*, [online] last access: 25.01.2011; available at: http://www.dhb.de/index.php?id=42

DHB (2009b), *Finanz und Gebührenordnung*, [online] last access: 26.01.2011; available at: http://www.dhb.de/index.php?id=42

DISTASO W., LEONIDA L., PATTI D. M. A. AND NAVARRA P. (2008), Corruption and Referee Bias in Football: The Case of Calciopoli, Italian Society of Public Economics, Department of Public and Territorial Economics, University of Pavia

DIXIT A. K. (1987), "Strategic Behavior in Contests" in American Economic Review 77, 891-898

DOHMEN T. J. (2008), "The Influence of Social Forces: Evidence from the Behavior of Football Referees" in Economic Inquiry 46(3), 411-424

DPA (2008), NBA-Schiedsrichter muss ins Gefängnis, [online] last access: 12.11.2009; available at: http://www.welt.de/sport/article2262584/ NBA-Schiedsrichter-muss-ins-Gefaengnis.html

DPA (2009a), Bestechungsvorwürfe: Handballbund suspendiert umstrittene Referees, [online] last access: 11.11.2009; available at: http://www.stern.de/sport/sportwelt/bestechungsvorwuerfe-handball-bund-suspendiert-umstrittene-referees-658052.html

DPA (2009b), Fußball-Wettskandal: Mehr als 30 Spiele in Deutschland unter Manipulationsverdacht, [online] last access: 01.11.2010; available at: http://www.spiegel.de/sport/fussball/0,1518,662444,00.html

DUYAR Z. (2009), Schiedsrichterbestechung und §263 StGB, Dissertation, Juristische Fakultät, Universität Bielefeld

EBERT J. (1980), Olympia: Mythos und Geschichte Moderner Wettkämpfe, 1st Edition, Wien: Edition Tusch

EHRENBERG R. G. AND BOGNANNO M. L. (1990), "Do tournaments have incentive effects" in Journal of Political Economy 98, 1307-1324

EHRLICH I. (1996), "Crime, Punishment and the Market of Offenses" in *Journal of Economic Perspectives 10(1)*, 43-67

EISENHARDT K. M. (1985), "Control: Organizational and Economic Approaches" in *Management Science 31*, 134-149

EISENHARDT K. M. (1989), "Agency Theory: An Assessment and Review" in *Academy of Management Review 14*, 57-74

EPSTEIN G.S. AND HEFEKER C. (2003), "Lobbying Contests with Alternative Instruments" in *Economics of Governance 4*, 81-89

FAMA E. F. (1980), "Agency Problems and the Theory of the Firm" in *Journal of Political Economy 88*, 288-307

FAMA E. F. AND JENSEN M. C. (1983a), "Separation of Ownership and Control" in *Journal of Law and Economics 26*, 301-325

FAMA E. F. AND JENSEN M. C. (1983b), "Agency Problems and Residual Claims" in *Journal of Law and Economics 26*, 326-349

FARMER A. AND PECORINO P. (1999), "Legal expenditure as a rent-seeking game" in *Public Choice 100*, 271-288

FINLEY M. I. AND PLEKET H. (1976), *Die Olympischen Spiele der Antike*, Tübingen: Rainer Wunderlich Verlag Hermann Leins

FOCUS (2009), *Schiedsrichter-Bestechung: Einladung zum Betrug*, [online] last access: 26.01.2011; available at: http://www.focus.de/sport/mehrsport/schiedsrichter-bestechung-einladung-zum-betrug_aid_380524.html

FRICK B. (2003), "Contest Theory and Sport" in *Oxford Review of Economic Policy 19(4)*, 512-529

FRITZWEILER J. (1998), "Ein §299a StGB als neuer Straftatbestand für den sich dopenden Sportler?" in *Zeitschrift für Sport und Recht*, 234-235

FUDENBERG D., GIBLERT R. AND TIROLE J. (1983), "Preemption, Leapfrogging and Competition in Patent Races" in *European Economic Review 22*, 3-32

GARICANO L., PALACIOS-HUERTA I. AND PRENDERGAST C. (2005), "Favoritism under social pressure" in *Review of Economics and Statistics 87*, 208-216

GLAZER A. AND HASSIN R. (1988), "Optimal Contests" in *Economic Inquiry 26(1)*, 133-143

GÖDECKE C. (2010), *DFB Sieg gegen England – Löws Weltmeisterprüfung*, [online] last access: 29.06.2010; available at: http://www.spiegel.de/sport/fussball/0,1518,703205,00.html

GOEL R. K. AND RICH D. P. (1989), "On the Economic Incentives for Taking Bribes" in *Public Choice 61(3)*, 269-275

HANDELSBLATT (2005), *Vereinsaustritt: Hoyzer entzieht sich DFB-Bestrafung*, [online] last access: 17.01.2011; available at: http://www.handelsblatt.com/sport/sonstiges/hoyzer-entzieht-sich-dfb-be-strafung;884274

HARBRING C. AND IRLENBUSCH B. (2008), "How many winners are good to have? On tournaments with sabotage" in *Journal of Economic Behavior and Organization 65(3-4)*, 682-702

HARBRING C., IRLENBUSCH B., KRÄKEL M. AND SELTEN R. (2004), *Sabotage in Asymmetric Contests*, mimeo, University of Cologne and London School of Economics

HARRIS C. AND VICKERS J. (1985), "Patent Races and the Persistence of Monopoly" in *Journal of Industrial Economics 33*, 461-481

HEERMANN P. H. (2009), "Zivilrechtliche Haftung für Fehlverhalten des Schiedsrichters" in Württembergischer Fußballverband e.V. (ed.), *Der Schiedsrichter im Spannungsfeld zwischen Anforderung und Überforderung*, 1st Edition, Nomos: Baden-Baden, 45-75

HELLMUTH I. AND EWERS C. (2009), *Handball Funktionär Butzeck: "Unser Präsident ist untragbar"*, [online] last access: 01.11.2010; available at: http://www.stern.de/sport/sportwelt/handball-funktionaer-butzeck-unser-praesident-ist-untragbar-658433.html

HILLMAN A. AND RILEY J. G. (1989), "Politically Contestable Rents and Transfers" in *Economics and Politics 1*, 17-40

HUBA K.-H. (2007), *Fußball Weltgeschichte: 1846 bis heute; Bilder, Daten, Fakten*, München: Copress Sport

JAIN A. K. (2001), "Corruption: A Review" in *Journal of Economic Surveys 15*, 71-121

JÄGER C. (2010), "Die drei Unmittelbarkeitsprinzipien beim Betrug" in *Juristische Schulung*, 761-766

JENSEN M. C. AND MECKLING W. H. (1976), "Theory of the Firm: Managerial Behaviour, Agency Costs and Ownership Structure" in *Journal of Financial Economics 3*, 305-360

JOST P.-J. (1999), *Strategisches Konfliktmanagement in Organisationen - Eine spieltheoretische Einführung*, 2nd Edition, Wiesbaden: Gabler Verlag

JOST P.-J. (2000a), *Ökonomische Organisationstheorie*, 1st Edition, Wiesbaden: Gabler Verlag

JOST P.-J. (2000b), *Organisation und Motivation - Eine ökonomisch-psychologische Einführung*, 1st Edition, Wiesbaden: Gabler Verlag

JUSTIZ BAYERN (2009), *Referentenentwurf - Gesetz zur Bekämpfung des Dopings und der Korruption im Sport*, [online] last access: 16.10.2010; available at: http://www.justiz.bayern.de/imperia/md/content/stmj_internet/ministerium/ministerium/gesetzgebung/entwurf_sportschutz-gesetz_30112009.pdf

KARGL W. (2007), "Begründungsprobleme des Dopingstrafrechts" in *Neue Zeitschrift für Strafrecht*, 489-496

KATZ A. (1988), "Judicial Decisionmaking and Litigation Expenditure" in *International Review of Law and Economics 8*, 127-143

KINGSTON C. (2007), "Parochial Corruption" in *Journal of Economic Behavior and Organization 63*, 73-87

KLEIBER D. A. AND ROBERTS G. C. (1981), "The effects of sport experience in the development of social character: A preliminary investigation" in *Journal of Sport and Exercise Psychology 3*, 114-122

KLIMKE B. (2005), *Korruption im Fußball gibt es strafrechtlich nicht*, [online] last access: 01.11.2010; available at: http://www.berlinonline.de/berliner-zeitung/archiv/.bin/dump.fcgi/2005/0204/sport/0008/index.html

KOCH R. AND MAENNIG W. (2007), "Spiel- und Wettmanipulationen - und der Anti-Korruptionskampf im Fußball" in *Der Bürger im Staat 56(1)*, 50-58

KÖNIG P. (2010), "Sportschutzgesetz - Pro und Contra; Pro: Argumente für ein Sportschutzgesetz" in *Zeitschrift für Sport und Recht*, 106-107

KONRAD K. A. (2000), "Sabotage in Rent Seeking Contests" in *Journal of Law, Economics and Organization 16*, 155-165

KONRAD K. A. (2009), *Strategy and Dynamics in Contests: London School of Economics Perspectives in Economic Analysis*, New York: Oxford University Press

KOOREMAN P. AND SCHOONBEEK L. (1997), "The Specification of the Probability Functions in Tullock's Rent-Seeking Contest" in *Economics Letters 56*, 59-61

KRACK R. (2007), *Zeitschrift für Internationale Strafrechtsdogmatik 3*, 103-112

KRÄKEL M. (2005), "Helping and Sabotaging in Tournaments" in *International Game Theory Review 7*, 211-228

KRÄKEL M. (2006), "Doping and Cheating in Contest-Like Situations" in *European Journal of Political Economy 23(4), 998-1006*

KRÄKEL M. AND SLIWKA D. (2004), "Risk taking in asymmetric tournaments" in *German Economic Review 5*, 103-116

KRÜGER A. O. (1974), "The Political Economy of the Rent-Seeking Society" in *American Economic Review 64*, 291-303

KUDLICH H. (2010), "Sportschutzgesetz - Pro und Contra; Contra: Argumente gegen ein Sportschutzgesetz" in *Zeitschrift für Sport und Recht*, 108-109

KUHN A. (2001), *Der Schiedsrichter zwischen bürgerlichem Recht und Verbandsrecht*, Frankfurt: Peter Lang

LAZEAR E. P. (1989), "Pay Equality and Industrial Politics" in *Journal of Political Economy 97*, 561-580

LAZEAR E. P. AND ROSEN S. (1981), "Rank-Order Tournaments as Optimum Labor Contracts" in *Journal of Political Economy 89*, 841-864

LEININGER W. (1991), "Patent Competition, Rent Dissipation and the Persistence of Monopoly" in *Journal of Economic Theory 53*, 146-172

LOURY G. (1979), "Market structure and innovation" in *Quarterly Journal of Economics 93*, 395-410

MAENNIG W. (2002), "On the Economics of Doping and Corruption in International Sports" in *Journal of Sports Economics 3*, 61-89

MAENNIG W. (2005), "Corruption in International Sports and Sport Management: Forms, Tendencies, Extent and Countermeasures" in *European Sport Management Quarterly 5(2)*, 187-225

MASON D. S., THIBAULT L. AND MISENER L. (2006), "An Agency Theory Perspective in Sport: The Case of the International Olympic Committee" in *Journal of Sport Management 20*, 52-73

MCCORMACK J. B. (1988), "Sport as Socialization: A Critique of Methodological Premises" in *The Social Science Journal 25(1)*, 83-92

MEZÖ F. (1980), *Die Geschichte der Olympischen Spiele*, München: Knorr & Hirth GmbH

MOLDOVANU B. AND SELA A. (2001), "The Optimal Allocation of Prizes in Contests" in *American Economic Review 91(3)*, 542-558

NALEBUFF B. AND STIGLITZ J. E. (1983), "Prizes and Incentives: Towards a General Theory of Compensation and Competition" in *Bell Journal of Economics 14*, 21-43

NEVILL A. M., BALMER N. J. AND WILLIAMS A. M. (2002), "The influence of crowd noise and experience upon refereeing decisions in football" in *Psychology of Sport and Exercise 3*, 261-272

NITZAN S. (1994), "Modelling Rent-Seeking Contests" in *European Journal of Political Economy 10*, 41-60

NTI K. (1999), "Rent-Seeking with Asymmetric Valuations" in *Public Choice 98*, 415-430

O'KEEFFE M., VISCUSI W. K. AND ZECKHAUSER R. J. (1984), "Economic Contests: Comparative Reward Schemes" in *Journal of Labor Economics 2*, 27-56

PEER M. (2009), *Korruption im Fußball nimmt zu*, [online] last access: 01.11.2010; available at: http://www.handelsblatt.com/sport/fussball/korruption-im-fussball-nimmt-zu;2464003

POSNER R. A. (1975), "The social costs of monopoly and regulation" in *Journal of Political Economy 83*, 807-827

PRANTL H. (2009), *Gesetzesentwurf aus Bayern: Zehn Jahre Haft für Doping und Sportbetrug*, [online] last access: 15.04.2010; available at: http://www.sueddeutsche.de/sport/312/495636/text/print.html

PRESTON I. AND SZYMANSKI S. (2003), "Cheating in Contests" in *Oxford Review of Economic Policy 19*, 612-624

RESCHKE J. AND KNAACK B. (2010), *4:1 Sieg – Deutschland müllert England weg*, [online] last access: 29.06.2010; available at: http://www.spiegel.de/sport/fussball/0,1518,703156,00.html

ROSE-ACKERMAN S. (1975), "The Economics of Corruption" in *Journal of Public Economics 4*, 187-203

ROSEN S. (1986), "Prizes and Incentives in Elimination Tournaments" in *American Economic Review 76*, 701-715

ROSS S. AND SZYMANSKI S. (2003), *The Law and Economics of Optimal Sports League Design*, Illinois Public Law Research Paper No. 03-14.

SAGE G. H. (1990), *Power and ideology in American sport: A critical perspective*, Champaign: Human Kinetics

SALIGER F., RÖNNAU T. AND KIRCH-HEIM C. (2007), "Täuschung und Vermögensschaden beim Sportwettbetrug durch Spielteilnehmer - Fall 'Hoyzer'", in *Neue Zeitschrift für Strafrecht*, 361-368

SCHLÖSSER J. (2005), "Der 'Bundesliga-Wettskandal' - Aspekte einer strafrechtlichen Bewertung" in *Neue Zeitschrift für Strafrecht*, 423-429

SCHÖNKE A. AND SCHRÖDER H. (2010), *Strafgesetzbuch*, 28th Edition, C. H. Beck

SCHREIBER H.-L. AND BEULKE W. (1977), "Untreue durch Verwendung von Vereinsgeldern zu Bestechungszwecken - BGH NJW 1975, 1234" in *Juristische Schulung*, 656-661

SCHULZE T. (2009), *HSV-Handball-Präsident Andreas Rudolph: "Der Schaden ist gewaltig"*, [online] last access: 26.01.2011; available at: http://www.stern.de/sport/sportwelt/hsv-handball-praesident-andreas-rudolph-der-schaden-ist-gewaltig-658122.html

SHLEIFER A. AND VISHNY R. W. (1993), "Corruption" in *The Quarterly Journal of Economics 108(3)*, 599-617

SKAPERDAS S. (1996), "Contest Success Functions" in *Economic Theory 7(2)*, 283-290

SKAPERDAS S. AND GROFMAN B. (1995), "Modeling Negative Campaigning" in *American Political Science Review 89(1)*, 49-61

SMIRRA N. (2008), *Doping, Regelbruch und Spielmanipulationen. Zur wettbewerbsrechtlichen Betrachtung des Profisports*, [online] last access: 23.04.2010; available at: http://www.ip-notiz.de/ip-expertennotizen-doping-regelbruch-und-spielmanipulationen-m-zur-wett-bewerbsrechtlichen-betrachtung- des-profisports/2008/08/18/

SOSA L. A. (2004), "Wages and other Determinants of Corruption" in *Review of Development Economics 8(4)*, 597-605

SPIEGEL (2005), *Fall Hoyzer – DFL will Frühwarnsystem einführen*, [online] last access: 26.01.2011; available at: http://www.spiegel.de/sport/fussball/0,1518,338289,00.html

STEIN W. E. (2002), "Asymmetric Rent-Seeking with More than Two Contestants" in *Public Choice 113*, 325-336

STEINKE C., BODMER A.-M. AND TOPHOVEN M. (2010), *Superspitze Fussballwitze*, Stuttgart: Franckh-Kosmos

STERN (2009), *Manipulationen im Handball: Hinweise auf flächendeckendes Bestechungssystem*, [online] last access: 01.11.2010; available at: http://www.stern.de/sport/sportwelt/manipulationen-im-handball-hinweise-auf-flaechendeckendes-bestechungssystem-658807.html

STIGLER G. J. (1970), "The Optimum Enforcement of Laws" in *Journal of Political Economy 78*, 526-536

SÜDDEUTSCHE ZEITUNG (2006), *Gesetzeslücken für Sportbetrüger*, [online] last access: 15.04.2010; available at: http://www.sueddeutsche.de/ sport/48/381850/test/

SUMMERER T. (1998), "Autonomie, Organisation, Regelwerk und Management des Sports" in Fritzweiler J., Pfister B. and Summerer T. (eds.), *Praxishandbuch Sport*, 1st Edition, C. H. Beck, 104-153

SUTTER M. AND KOCHER M. (2004), "Favoritism of Agents – The Case of Referees' Home Bias" in *Journal of Economic Psychology 25*, 461-469

SZYMANSKI S. (2003a), "The Assessment: The Economics of Sport" in *Oxford Review of Economic Policy 19(4)*, 467-477

SZYMANSKI S. (2003b), "The Economic Design of Sporting Contests" in *Journal of Economic Literature 41*, 1137-1187

SZYMANSKI S. AND VALLETTI T. M. (2005), "Incentive Effects of Second Prizes" in *European Journal of Political Economy 21(2)*, 467-481

TAYLOR B. A. AND TROGDON J. G. (2002), "Losing to win: Tournament incentives in the National Basketball Association" in *Journal of Labor Economics 20*, 23-41

TIROLE J. (1986), "Hierarchies and Bureaucracies: On the Role of Collusion in Organizations" in *Journal of Law, Economics and Organization 2(2)*, 181-214

TRIFFTERER O. (1975), "Vermögensdelikte im Bundesligaskandal" in *Neue Juristische Wochenschrift*, 612-617

TULLOCK G. (1967), "The Welfare Cost of Tariffs, Monopolies, and Theft" in *Western Economic Journal 5*, 224-232

TULLOCK G. (1980), "Efficient Rent-Seeking" in Buchanan J. M., Tollison R. D. and Tullock G. (eds.), *Towards a Theory of the Rent-Seeking Society*, College Station, Texas A&M University Press, 97-112

TUNEKE H. F. (2009), *"Bestechung kommt in allen Sportarten vor"*, [online] last access: 01.11.2010; available at: http://www.derwesten.de/sport/basketball/Bestechung-kommt-in-allen-Sportarten-vor-id567000.html

TURNER G. (1992), "Ist ein Anti-Doping-Gesetz erforderlich?" in *Zeitschrift für Rechtspolitik*, 121

WABNITZ H.-B. AND JANOVSKY T. (2007), *Handbuch des Wirtschafts- und Steuerstrafrechts*, 3rd Edition, C.H. Beck

WEGMANN H. (2010), "Entwurf zum Sportschutzgesetz: ja - aber richtig" in *Causa Sport 3*, 242-246

WEINREICH J. (2009), *Korruption im Handball: Der Aufstieg des Manipulators*, [online] last access: 01.11.2010; available at: http://www.stern.de/sport/sportwelt/korruption-im-handball-der-aufstieg-des-manipulators-658848.html

VAN RIJCKEGHEM C. AND WEDER B. (2001), "Bureaucratic Corruption and the Rate of Temptation: Do Wages in the Civil Service Affect Corruption, and by How Much?" in *Journal of Development Economics 65*, 307-331

VOIGT A. (2009), *Basketball-Schiris: "Man kann Bestechung nie ausschließen"*, [online] last access: 26.01.2011; available at: http://www.spiegel.de/sport/sonst/0,1518,623733,00.html

VOLKERY C. (2010), *Englisches Disaster – 'Ich halte das nicht mehr aus*, [online] last access: 29.06.2010; available at: http://www.spiegel.de/panorama/0,1518,703193,00.html

VON KOMOROWSKI A. AND BREDEMEIER B. (2005), "Fußball, Vermögensstrafrecht, und Schiedsrichterverhalten" in *Zeitschrift für Sport und Recht*, 181-184

WARREN D. E. AND LAUFER W. S. (2009), "Are Corruption Indices a Self-Fulfiling Prophecy? A Social Labeling Perspective of Corruption" in *Journal of Business Ethics 88(4)*, 841-849

ZAUGG K. (2010), *Darum wird es immer Korruption geben*, [online] last access: 26.10.2010; available at: http://i.20min.ch/de/sport/28558315/Darum-wird-es-immer-Korruption-geben

ZIEHER W. (2009), "Fehlverhalten des Schiedsrichters aus strafrechtlicer Sicht" in Württembergischer Fußballverband e.V. (ed.), *Der Schiedsrichter im Spannungsfeld zwischen Anforderung und Überforderung*, 1st Edition, Nomos: Baden-Baden, 25-44

GABLER RESEARCH

„Management, Organisation und ökonomische Analyse"
Herausgeber: Prof. Dr. Peter-J. Jost
zuletzt erschienen:

Band 10
Clemens Löffler
Strategische Selbstbindung und die Auswirkung von Zeitführerschaft
2008. XV, 205 S. ,18 Abb., 34 Tab., Br. € 59,95
ISBN 978-3-8349-1178-0

Band 11
Bernd-Oliver Heine
Konzeptionelle Nutzung von Controllinginformationen
Ein modelltheoretischer Ansatz
2008. XXI, 237 S., 24 Abb., 27 Tab., Br. € 59,95
ISBN 978-3-8349-1188-9

Band 12
Julia Hornstein
Modellgestütze Optimierung des Führungsstils während eines Turnarounds
2009. 224 S., 53 Abb., Br. € 59,95
ISBN 978-3-8349-1731-7

Band 13
Holmer Glietz
Ökonomische Analyse des mittleren Managements
Organisationsstrukturen und Reorganisationen
2011. XIII, 296 S., 20 Abb., Br. € 59,95
ISBN 978-3-8350-0164-0

Band 14
Cedric Duvinage
Referees in Sports Contests
Their Economic Role and the Problem of Corruption
in Professional German Sports Leagues
2012. XVI, 161 S., 2 Abb., 1 Tab., Br. € 64,15
ISBN 978-3-8349-3526-7

Änderungen vorbehalten. Stand: November 2011.
Erhältlich im Buchhandel oder beim Verlag.
Gabler Verlag . Abraham-Lincoln-Str. 46 . 65189 Wiesbaden . www.gabler.de

GABLER